ETHICS AND MENTAL RETARDATION

PHILOSOPHY AND MEDICINE

Editors:

H. TRISTRAM ENGELHARDT, JR.

The Center for Ethics, Medicine and Public Issues, Baylor College of Medicine, Houston, Texas, U.S.A.

STUART F. SPICKER

University of Connecticut, School of Medicine, Farmington, Connecticut, U.S.A.

VOLUME 15

ETHICS AND MENTAL RETARDATION

Edited by

LORETTA KOPELMAN

and

JOHN C. MOSKOP

East Carolina University School of Medicine, Greenville, N.C., U.S.A.

D. REIDEL PUBLISHING COMPANY

A MEMBER OF THE KLUWER ACADEMIC PUBLISHERS GROUP

DORDRECHT / BOSTON / LANCASTER

Library of Congress Cataloging in Publication Data

Main entry under title:

Ethics and mental retardation.

(Philosophy and medicine ; v. 15)
Includes index.
1. Mental retardation—Moral and ethical aspects. 2. Mental retardation—Religious aspects. 3. Mental health laws. I. Kopelman, Loretta M., 1938- . II. Moskop, John C., 1951- . III. Series.
[DNLM: 1. Mental retardation–Congresses. 2. Mental retardation–Legislation and jurisprudence–Congresses. 3. Ethics–Congresses. 4. Public policy–Congresses. 5. Human rights–Congresses.
W3 PH609 v. 15]
RC570.E84 1984 179'.7 83-24561
ISBN 90-277-1630-7

Published by D. Reidel Publishing Company,
P.O. Box 17, 3300 AA Dordrecht, Holland.

Sold and distributed in the U.S.A. and Canada
by Kluwer Academic Publishers,
190 Old Derby Street, Hingham, MA 02043, U.S.A.

In all other countries, sold and distributed
by Kluwer Academic Publishers Group,
P.O. Box 322, 3300 AH Dordrecht, Holland.

Printed in The Netherlands

For Arthur and Ruth

TABLE OF CONTENTS

SECTION IV / LAW AND PUBLIC POLICY

ACKNOWLEDGEMENTS

To the East Carolina University School of Medicine and the North Carolina Humanities Committee we express our deep appreciation for sponsoring the symposium which became the basis for this book. The symposium, entitled "Natural Abilities and Perceived Worth: Rights, Values and Retarded Persons", was held on October 1–3, 1981, in Greenville, North Carolina. We are grateful to the faculty of the School of Medicine and of the Department of Philosophy at East Carolina University for their support of this project. In particular we want to thank William Laupus, M.D., Dean of the School of Medicine, Thomas B. Brewer, former Chancellor of East Carolina University, R. Oakley Winters, former executive director of the North Carolina Humanities Committee, and the general editors of this series, H. Tristram Engelhardt, Jr. and Stuart F. Spicker, for their encouragement and assistance. A special debt of gratitude is owed to our contributors for their willingness to explore a subject that is doubly difficult because it is both uncharted and interdisciplinary. We wish to thank Robert M. Adams and Jeffrie Murphy for allowing us to include their previously published papers in this volume. (Adam's 'Must God Create the Best?' originally appeared in the *Philosophical Review,* **81**, 1972, and Murphy's 'Rights and Borderline Cases' appeared in the *Arizona Law Review*, **19**, 1977.) Many challenging and tedious tasks in preparing for the symposium and volume fell to our secretaries Diane Greer, Ramona Shannon and Joanne Stoddard. We thank them for their patience, care and dedication in working with us.

LORETTA KOPELMAN
JOHN MOSKOP

INTRODUCTION

This volume offers a collection of writings on ethical issues regarding retarded persons. Because this important subject has been generally omitted from formal discussions of ethics, there is a great deal which needs to be addressed in a theoretical and critical way. Of course, many people have been very concerned with practical matters concerning the care of retarded persons such as what liberties, entitlements or advocacy they should have. Interestingly, because so much practical attention has been given to issues which are not discussed by ethical theorists, they offer a rare opportunity to evaluate ethical theories themselves. That is, certain theories which appear convincing on other subjects seem implausible when they are applied to reasoned and compelling views we hold concerning retarded individuals.

Our subject, then, has both practical and conceptual dimensions. Moreover, because it is one where pertinent information comes from many sources, contributors to this volume represent many fields, including philosophy, religion, history, law and medicine. We regret that it was not possible to include more points of view, like those of psychologists, sociologists, nurses and families. There is however, a good and longstanding literature on mental retardation from these perspectives. We sought, rather, to address a neglected topic: How can we identify and evaluate the assumptions in our laws, religious, morals and mores which shape our judgments about retarded individuals? Are we kind and fair to them in the way we frame our theories and our laws?

It would be easier and certainly less embarrassing to identify unwarranted assumptions of peoples or ages other than our own. It is not difficult, for example, to ridicule pretentious "scientific" claims about the retarded made by the "experts" of the early part of this century. Though they claimed to *know* that most mental retardation is hereditary along Mendelian lines, it is now held that less than 20% of it is. They claimed to *know*, moreover, that mental retardation is closely associated with deviant behavior, including crime, alcoholism and sexual promiscuity. The pompous trappings of their sweeping and unfounded claims might seem amusing were it not for the harm done by them. Whether people found the "facts" they wanted to find or were sincerely misled by faulty reasoning, it was too easy a step from this hard "data" to eugenics movements with laws requiring sterilization and other restrictions of the freedoms of citizens whom society called 'retarded'. These movements, invoking the authority of science, encouraged the cruel devaluation of people who, through no fault of their own, were retarded or who had retarded relatives. Whatever burdens these people had were increased rather than diminished by society. It is not difficult to find the remnants of these views in our society. We believe that attention to the moral issues raised by retarded individuals will help us identify and overcome unwarranted attitudes and practices.

The manner in which retarded members of our communities are viewed and treated, then, not only makes a difference in their lives but also tells us a good deal about ourselves. Some of the more practical issues (and there has been lively debate on these for some time) are: How should we allocate resources to help retarded citizens and their families? How well have institutions worked and what alternatives should be available? What are the motives for depopulation of our institutions: kindness, economics or expediency? On what basis do we ascribe handicaps? More theoretical questions can also be raised: What kinds of values are entailed by the ascription of handicaps to others? How do the rights shown retarded citizens square with current moral theories? These questions, addressed by our contributors, have important implications for our traditions and ourselves. They are difficult and uncomfortable questions, since providing better care and opportunities makes demands, even on an affluent society.

Section I of the volume emphasizes the application of theories of rights and justice to retarded persons. Jeffrie Murphy in his "Rights and Borderline Cases" argues that autonomy rights must be supplemented by a notion of social contract rights, and to do this he draws on John Rawls' theory of justice. Since autonomy rights presuppose agency, profoundly retarded persons could not exercise such rights. Murphy argues, however, that it is reasonable to establish certain minimum levels of security below which no one should fall. Social contract rights reasonably should be provided to those who have special needs, like infants or retarded citizens, for our sakes (to preserve our just institutions) as well as theirs (out of concern for what they feel). In contrast to Murphy, Joseph Margolis argues that no general doctrine of human rights or justice can convincingly be used to give specific shape to the rights of the retarded or any other group. All such general doctrines, Margolis argues, face the following dilemma: Either the principles are so abstract that no specific claims can be drawn from them, or, if specific claims can be derived, they may be incompatible with the actual resources or limitations of particular societies. Margolis himself prefers to ground special treatment for retarded persons on liberal principles actually held by particular societies. In a rejoinder Murphy ironically entitles "Do the Retarded Have a Right Not to be Eaten," he charges that without an appeal to rights, Margolis' "liberal principles" cannot justify a distinction between retarded individuals and non-human animals. Despite difficulties with theories of rights, then, Murphy still finds "rights talk" preferable to Margolis' alternative.

In his commentary, Tristram Engelhardt underlines Margolis' admission that considerations of consistency can establish strong rights to procedural fairness. He also generalizes Margolis' criticisms of Rawls to all hypothetical choice theories of morality; all such theories, he claims, presuppose substantial moral issues in framing the hypothetical choice situation. He concludes that although we rely on moral traditions to illustrate different values and practices, the ultimate choice among them is not dictated by overarching principles.

Anthony Woozley and Cora Diamond address a crucial question: Is our treatment of the retarded a matter of *justice* or of *charity*? In discussing this they, like Margolis, question the adequacy of the rights tradition. The preoccupation with rights, they argue, may cause us to neglect compelling needs of retarded persons as unique individuals. Woozley distinguishes between our regard for someone as an individual and as a member of some group each of whose members must be treated in a certain way.

There are dangers in becoming so preoccupied with rights that good sense and charity are ignored. In her commentary, Cora Diamond cites examples of attitudes towards retarded persons in 19th-century Russia and the Indian subcontinent where treatment of the retarded was motivated by care and compassion, not out of regard for rights. Diamond concludes that the requirements of justice and charity are closely linked.

In Section II Loretta Kopelman defends the view that all humans merit respect. But 'respect' cannot be understood here as 'esteem' or require any individual features (such as intelligence, or moral personality). Rather, we respect their status as fellow-beings for their sakes as well as ours. This kind of respect is illustrated by the moral obligation to justify ascriptions of handicaps. Fulfilling this requirement shows a respect we think rational whatever the capacity of the individual evaluated. Like Woozley, she criticizes Rawls' discussion of the retarded. She argues that Rawls takes respect for persons as basic and fails to give a plausible account of why the severely retarded are rights-bearers. In his commentary, Stuart Spicker agrees that interpersonal relationships are crucial, but argues that an attitude of respect is always tied to a concept of reciprocal personal agency. Spicker distinguishes between a principle of respect and an attitude of respect, where the attitude of respect requires agency and reciprocity. Though severely retarded individuals cannot coherently claim a right to our attitude of respect, Spicker argues, one can act from a principle of respect and thus it may be argued that we have obligations to them nonetheless.

Laurence McCullough questions whether special treatment for the retarded must be purchased at the costly price of a general loss of independence and respect for them as persons. He argues that this price need not be paid if we replace a "fixed state" definition of handicaps for a developmental or "process oriented" model, and if we determine competency less globally as related to specific tasks. By more clearly delineating kinds of disability included in mental retardation, McCullough claims, we can justify provision of special treatment without wide-ranging loss of moral status. In commenting on this essay, Robert Holmes questions whether McCullough has, as he claims, found a conceptual muddle at the heart of the definition of mental retardation. Holmes also argues that despite McCullough's concern to defend the interests of retarded persons, he has, in fact, committed himself to the mistaken view that diminished moral status (that is, loss of rights and responsibilities) is associated with mental retardation.

Section III of the volume explores issues in theology and philosophy of religion. Robert M. Adams, in "Must God Create the Best?", confronts in a theoretical way a question that occurs to many families who first learn they have a retarded child. A first response of many is angrily to question why God has permitted this to occur. Adams *rejects* the principle, "If a perfectly good moral agent created any world at all, he would have to create the best world that he could create." Adams argues that from the Judaeo-Christian standpoint, creation of a less excellent world is consistent with perfect goodness, since it can be explained in terms of God's grace in freely choosing to love creatures who cannot merit his love.

William May emphasizes an interpersonal or relational account of valuing retarded persons which is grounded in the significance of parenting and bonding. May rejects easy sentimentality in exploring the nature of "the ordeal" of detachment, transition, and attachment to a retarded child. His rich descriptions support his view that "acceptance", a word popular in the literature on retarded children, is too simplistic a notion to capture the complexities and ambiguities of the process. Praising May for the power and persuasiveness of his essay, commentator Larry Churchill calls it a portrait of, rather than an argument for, a position. This is meant less as a criticism than as a call for further study, elaboration and justification. Churchill applauds May's effort to bring the experience of parents in caring for retarded children to the forefront of discussions about valuing the retarded.

John Moskop reviews approaches to retarded persons offered by theologians Joseph Fletcher and Stanley Hauerwas. Moskop criticizes Fletcher's singleminded concern to prevent retardation for its insensitivity to the worth and needs of present and future retarded persons. Though more sympathetic to Hauerwas' conception of retarded children as gifts from God, Moskop is concerned about a question Hauerwas does not discuss, namely, "Why does God create retarded persons?" Reflection on this question leads him to suggest that God's special concern for retarded children can be affirmed without asserting that God wills that they be retarded. In response, Churchill claims that Moskop imposes the philosophical problem of evil on a foreign theological context. To support this claim, Churchill presents other approaches to this problem which he finds more congenial to Hauerwas' own confessional point of view.

In Section IV, Michael Kindred reviews the legal treatment of mentally retarded persons during this century. He identifies three periods marked by different legal approaches. In the first, lasting well into the 1960's, legal rights for retarded persons were generally neglected by the courts. In the second period, the 1970's, the courts examined issues of due process and equal protection for the retarded. As part of civil rights reform, courts brought recognition of significant rights for the retarded in the areas of education, involuntary institutionalization and habilitation. More recent and restrictive rulings by the U.S. Supreme Court, however, suggest to Kindred that a third period has begun. These rulings, he argues, constitute a retreat from the sweeping reforms and regulations introduced in the 1970's.

Barbara Levenbook examines possible philosophical justifications for restricting the legal rights of retarded persons, especially rights to freedom and self-determination. Though they have been frequently used, she argues that principles of paternalism and distributive justice are probably not adequate to justify such legal restrictions.

David Rothman raises the fundamental question of who is best qualified to represent those more severely retarded persons unable to make their own decisions. Rothman argues that professional care-givers, families and lawyers have other interests which may interfere with advocacy for the retarded and concludes that special advocates are the best representatives for retarded persons. He is skeptical about relying too heavily on the good will of individuals and stresses the importance of entitlements to minimum standards.

In his commentary, psychiatrist-neurologist Gerald Moriarty illustrates some of the different kinds of impairments suffered by retarded persons and notes how these interfere with their ability to control their lives. Moriarty is sceptical of the long-term value of court-ordered plans for the care of retarded persons. In place of adversarial processes, he stresses an atmosphere of co-operation among all those who provide services to the retarded.

In the final essays of the volume, two physicians, Arthur Kopelman and Theodore Kushnick, raise urgent practical questions regarding the care of retarded persons. Dr. Kopelman points out financial obstacles to providing intensive medical care to seriously ill newborns. Dr. Kushnick stresses the importance of liberal diagnostic testing for the sake of handicapped infants and their families and also to prevent future tragedies. Both physicians lament the present lack of adequate medical and educational support services for the great majority of retarded persons.

The essays in this volume thus explore conceptual and philosophical issues of language, attitudes, policies and practices regarding mentally retarded persons. A focus on the relationship between ethical theory and social practices has allowed our contributors to evaluate both theory and practice. We wish to thank them for exploring a largely uncharted area of ethics.

February, 1983

LORETTA KOPELMAN
JOHN C. MOSKOP

SECTION I

EXAMINING THE RIGHTS TRADITION

JEFFRIE MURPHY

RIGHTS AND BORDERLINE CASES [1]

I. INTRODUCTION

Two competing theories of justice are currently dominating discussion within contemporary Anglo-American moral, social, political, and legal philosophy. The Welfare-State Liberal theory of John Rawls, developed most fully in his book *A Theory of Justice* [13], contains a sustained defense of the partial redistribution of wealth within society in order to improve the position of the most unfortunate members of society. The Libertarian theory of Robert Nozick, presented in his book *Anarchy, State, and Utopia* [12], is based on the claims that justice is the respecting of people's *rights*, that people's rights are respected when they are allowed to keep and control that to which they are *entitled,* and that people are entitled to any holding which they have legitimately acquired – that is, without using force, theft, or fraud (activities which would violate the rights of others). Since the kind of redistribution called for by Rawls' theory of justice would necessarily involve taking from some people (primarily through taxation) holdings to which they have a right, Nozick argues that an application of Rawls' theory will have unjust consequences: it will fail to show the proper respect for people's rights upon which any correct theory of justice must be based.

The present essay is an exploration of the concept of a *right* – a concept which is, as Nozick claims, central to the concept of justice. The essay is based on the assumption that one way to develop an understanding of this important moral concept is to examine cases – called 'borderline' here – where the ascription of rights seems doubtful or problematic. If we can understand what goes wrong in the doubtful cases, then perhaps we can develop a more satisfactory account of the clear cases. Although not written with such a purpose in mind, the essay may also be viewed as a tentative start toward showing that a generally Rawlsian theory of justice or rights can make an important place for those moral claims which Nozick calls entitlements.

L. Kopelman and J. C. Moskop (eds.), Ethics and Mental Retardation, 3–17.

II. THE PROBLEM

The nature and importance of the problem to be discussed in this essay has been well stated by Joel Feinberg:

> In the familiar cases of rights, the claimant is a competent adult human being, and the claimee is an officeholder in an institution or else a private individual, in either case, another competent adult human being. Normal adult human beings, then, are obviously the sorts of beings of whom rights can meaningfully be predicated. Everyone would agree to that .... On the other hand, it is absurd to say that rocks can have rights, not because rocks are morally inferior things unworthy of rights (that statement makes no sense either), but because rocks belong to a category of entities of whom rights cannot be meaningfully predicated. That is not to say that there are no circumstances in which we ought to treat rocks carefully, but only that the rocks themselves cannot validly claim good treatment from us. In between the clear cases of rocks and normal human beings, however, is a spectrum of less obvious cases, including some bewildering borderline ones. Is it meaningful or conceptually possible to ascribe rights to our dead ancestors? to individual animals? to whole species of animals? to plants? to idiots and madmen? to fetuses? to generations yet unborn? Until we know how to settle these puzzling cases, we cannot claim fully to grasp the concept of a right, or to know the shape of its logical boundaries. ([3], p. 44)

Feinberg's point can be summarized in this manner: We all recognize clear cases of individuals (competent adult human beings) to whom the ascription of rights, both moral and legal, is nonproblematic – indeed, it is obviously justified. We also recognize clear cases (such as rocks) of entities which not only do not have rights but *could not* have rights, such ascription being conceptually ludicrous. In addition to these clear cases, we also are confronted with a large class of borderline cases – individuals who have *some* of the features normally found in individuals who are clearly rights-bearers, but who do not seem to have *quite enough* of such features to make us confident in ascribing rights to them. They are of such a nature (for example, severely retarded) that they lack certain features (such as the power of rational choice) normally found in rights-bearers. Indeed they may even be of such a nature (for instance, irreversibly comatose) that they seem disturbingly similar in relevant respects to those entities (such as rocks) which clearly cannot be rights-bearers. All these considerations tempt us to say that they are *not* rights-bearers at all.

But there are temptations in the other direction as well. We are inclined to say that even the severely retarded, for example, may not be treated just *any* way we feel like – that a certain kind of decent treatment is *owed* to them, just as a certain kind of decent treatment is *owed* to animals and the insane. I use the term 'owed' at this point merely to indicate that we feel that the decent treatment is done at least in part *for the sake of these creatures* and not simply for our own sakes – to protect our sensitive feelings with respect to the hardships of other creatures – and that the decent treatment is *obligatory* on us and is not optional or supererogatory in the way that certain acts of kindness or beneficence are. For example, the feeling that we owe the severely retarded something and that any decent treatment we give them is not just a matter of charity is behind the temptation to say that even creatures such as these borderline ones *do* have rights. Perhaps they do not have as many rights as properly ascribable to competent adult humans, but they have some rights nevertheless. It is my hope in what follows to make a start toward sorting out our conflicting temptations here and to suggest a way, both conceptually clear and morally adequate, in which both sides can be satisfied – at least partially.

III. THE NATURE AND IMPORTANCE OF RIGHTS

I shall argue that we use the language of rights to do at least *two different* (though related) moral jobs, and that a failure to keep clear the distinction between these two jobs is behind a substantial amount of the fruitless philosophical controversy surrounding the question of whether borderline cases do or do not have rights.

One important function of the language of rights, stressed by such philosophers as John Locke and especially Immanuel Kant, has been to mark out the special kind of treatment (called "respect" by Kant) which is particularly fitting or appropriate to *autonomous, rational persons* (see [6,7,8]). There does seem to be something special about such persons (Kant calls it their "dignity") in that their destinies, at least in large measure, can be rationally controlled by choices or decisions. They are, to use another Kantian phrase, self-legislating members of a kingdom of ends (a community of other self-legislators).

In the spirit of Kant, Robert Nozick uses the phrase "rational creature" as "short for beings having those properties in virtue of which a being has those full rights that human beings now have" ([12], p. 299). This special feature of persons seems to provide a ground for a special set of moral claims (natural or human rights) which are appropriate only for such creatures and not for any other animals. These basic natural or human rights are articulated by Locke as follows: "No one ought to harm another in his life, health, liberty or possessions" ([8], section 6). This list is *negative* (it prohibits certain kinds of *interferences*) because only in this way can the special status of persons be protected and respected – their status as creatures who must be left their own "moral space" to work out their own destinies according to their own choices and decisions. Positive interference with them, even if benevolently or paternalistically motivated, would be an affront to their dignity or status as persons. Persons, Kant argued, are ends and not means only; they are not to be sacrificed or used as instruments or resources for the achieving of others' ends without their consent. They are inviolable. Stressing this inviolable nature of persons has been the central focus of many who have written on rights and – for what I hope are obvious reasons – I shall label their accounts of rights as accounts of *Autonomy Rights.*

Those like Kant who have stressed Autonomy Rights have not regarded such rights claims as being the only important part of morality, but they have certainly regarded them as the *most important* part. Kant argues that those duties resting on rights are absolute and decisive (he calls them "perfect") when in conflict with any other moral considerations – for example, when in conflict with such "imperfect" duties as beneficence (helping others in need). These imperfect duties do not rest on rights (I do not violate another's rights simply by refusing to help him), admit of "latitude in time and manner of their fulfillment" (I must simple do *something* for *some* people *some* of the time) and never take precedence over claims of genuine rights. *Rights claims are morally basic because the value of autonomy which they capture is evaluatively basic.* This, in very general and oversimplified terms, is the view of such writers as Kant and Nozick.

Another quite different tradition in moral and social philosophy, best represented by John Stuart Mill and our contemporary John Rawls, involves the view that rights claims are *not* morally basic. Rather, they are *derivative* from more general moral principles. "To have a right," Mill suggests, "is to have something which society ought to defend me in the possession of" ([9], p. 66).[2] The central idea here is that rights claims function, not to mark some specially fine feature of persons, but rather to mark out which of all moral claims *ought to be enforced by the state;* in other words, which ones *ought to be law.* On this view, a creature may be said to have a right to *x* if and only if it is morally reasonable, all things considered, to *guarantee x* to that creature as a matter of law.

Mill's own version of rights theory meets shipwreck in the answer he gives to the question: "How do we decide what is morally reasonable?" Mill, alas, is a utilitarian – one who claims that the basic or most important moral requirement is to maximize human happiness. This theory is so obviously morally bankrupt[3] that very few contemporary moral philosophers take it at all seriously. I will not take time here to reiterate all the things wrong with the theory. Suffice it to say, as a summary objection, that utilitarianism fails to pay attention to those important *autonomy* values discussed above and thus fails to articulate a satisfactory conception of *justice* or *respect for persons.* It does not, as any correct theory of justice would, rule out the sacrifice of persons for the general good.[4]

Just because Mill's ultimate moral theory is defective, however, it does not follow that his analysis of rights as claims that society ought to enforce is defective. Though John Rawls (a contemporary anti-utilitarian) says very little about rights, his analysis appears similar to that of Mill.[5] However, he does not give a utilitarian answer to the question of what is morally reasonable for society to enforce. Rather, he says we should ask ourselves this question: "If a group of ideally rational beings came together in order to pick rules to govern their mutual relations, which rules would they be compelled (by the power of their rationality) to pick?" This decision-theoretical interpretation of classical social contract theory is basic in Rawls' outlook.

Since I believe that, in spite of certain very serious problems,[6] Rawls' theory is very helpful in illuminating certain moral issues, I should like to discuss it in a bit more detail. As already indicated, Rawls suggests that we use the idea of a social contract as a model of rational decision in morality. That is, we shall regard as rationally justified any moral principle (or any social practice) if that principle (or practice) would be unanimously agreed to, adopted, or contracted for by a group of rational agents coming together, in what Rawls calls the "original position", in order to pick principles and then practices to govern their relations with each other as members of a common community. The "original position" is simply a hypothetical situation of priority to moral rules and social practices. Unanimous consent is required because it captures our intuitive notion of what is *fair* as an agreement (because this precludes anyone's being coerced), and this intuitive notion of fairness is prior to developed moral rules or principles. Indeed, according to the theory, such rules or principles will grow out of this intuitively acceptable setting. This is why Rawls speaks of justice *as* fairness. In an attempt to avoid begging any moral questions (for example, by smuggling certain controversial moral values into the very conception of rationality), Rawls stipulates that rational agents in the original position are primarily *self-interested* – primarily concerned to promote their own welfare (which includes promoting the welfare of those about whom they care deeply, such as members of their families). An important and controversial restriction on

Rawls' model of a rational being is that such a being is said to operate under a "veil of ignorance" ([13]), pp. 12, 19) – he is said to know what in general can happen to persons in various positions in society but not to know what *his* own actual position is likely to be. According to Rawls, this prevents the rational contractors from making choices on the basis of *morally irrelevant* considerations such as skin color. For example, in trying to decide whether or not to adopt a practice of slavery, each agent would know in general what it is like to be a slave, but none of the agents would know the probability of *his* being a slave.[7] Finally, Rawls suggests that each rational agent will value a set of primary goods (liberty, security, self-respect) and will always choose to minimize, as far as possible, threats to these goods. Rawls defines primary goods as "things that every rational man is presumed to want. These goods normally have a use whatever a person's rational plan of life" ([13], p. 62).

Given this brief sketch of an elaborate intellectual machinery, we are now in a position to see how Rawls attempts to show that rational persons would *not* choose utilitarianism as a basic moral principle. Since utilitarianism allows for the sacrifice of the few if this results in substantial benefit to the majority, it would never be unanimously agreed to in the original position. Rational agents, caring most about protecting such primary goods as liberty ("maxi-mining", in the economists' language), and having no guarantee (because of the veil of ignorance) that they will not be members of the sacrificed class, will choose a principle which does not allow such sacrifice. To return to the slavery example: Since no rational agent would be willing to run the risk of being a slave, rational agents would not unanimously agree to a principle which allowed slavery. As one can easily see, Rawls views rational people as *cautious* people – people who plan their lives on the least favorable set of assumptions about how things will turn out. Though they perhaps miss out on the bliss and exaltation of the gambler, they also (and this matters much more to them) miss out on the irrevocable disasters to which gamblers are prone.

What does all this have to do with rights? I will suggest the following as a Rawlsian-inspired analysis: An individual should be understood as having a right to x if and only if a law guaranteeing x to the individual would be chosen by rational agents in the original position. This concept of a right I shall call a *Social Contract Right.*

We have, then, *Autonomy Rights,* concerned to respect the choices and decisions of persons, and *Social Contract Rights,* concerned to provide morally justified legal guarantees. These are related, but a few moments of reflection should persuade us that they are not identical. If you promise that you will meet for dinner and then simply stand me up without good reason, you have shown insufficient respect for my humanity, treated me with contempt, and violated an autonomy right of mine. For surely it is part of the meaning of "I promise," sincerely uttered in the proper circumstances, that it gives a right to the person to whom the promise is made. But should all such rights be enforced by law? Surely not. Any state which sets for itself the goal of enforcing every promise, no matter how trivial, would be financially bankrupt and morally intrusive. This is why only certain promises (regarded as important enough to be made into *contracts*) are enforced by the state.

Consider another example: Suppose I am poor, unable to afford decent food and medical care. Does this mean that any of my autonomy rights have been violated? Not necessarily. Others may simply have exercised their right to leave me alone. They have neglected me but not interfered with me, and autonomy rights require only noninterference. And yet, many would want to say, the poor have a *right* to a decent standard of living, education, and health care. But why, if no considerations of autonomy are involved?[8] A Rawlsian answer here strikes me as plausible: Rational persons, in the original position, would want to protect themselves against certain major losses and harms. Thus they would probably agree to a *minimum floor* of security below which no member of society would be allowed to fall. Even selfish people, caring about their own welfare, would want to guard against the kind of destruction of primary goods that would result if they became severely disadvangaged, as people often do, through no fault of their own, but simply through their "bad luck on the social and natural lottery" ([13], p. 74).

In summary, most Autonomy Rights are going to be adopted as Social Contract Rights (though not all, as my promise example shows), because rational agents in the original position will surely want to protect themselves against interferences with their ability to determine their own future by their own choices. Social Contract Rights are going to be broader in scope than Autonomy Rights, however, because rational agents in the original position – in addition to caring about the protection of their autonomy against interference – are going to want to protect themselves against certain major hardships and sufferings. They know that they are sentient as well as sapient. They are thus likely to adopt minimum floors of security (income, health care and the like) and speak of these as *rights*. There will be neither confusion nor immorality here so long as they remain clear about the status of these rights – that they are Social Contract Rights and not Autonomy Rights. The main point in calling them rights is to emphasize that they are *guaranteed* – nonoptional.

What if autonomy as a value seems to compete with minimum floors as a value? Which will take precedence? In the original position, decisions will be made – not just about what rights to recognize – but about the priority of these rights claims. This is a difficult problem, but not, in my judgment, obviously insoluble.[9] At this point, I shall resist the strong temptation to consider in even greater detail the fascinating abstract issues raised by the Rawlsian outlook on rights, and return to the more practical issues with which I began.

IV. BORDERLINE CASES AGAIN

I shall not attempt to consider all of the borderline cases noted by Feinberg, but shall confine myself to a discussion of the extent to which, if at all *children* have rights, and the extent to which, if at all, *retarded individuals* have rights. I shall further confine my discussion to *very young* children and individuals who are *more than minimally* retarded. If we consider 17 year-old "children", or "retarded persons" with IQs of 85, it would seem unduly severe to treat them as any less than autonomous persons with the rights appropriate to that status. The actual location of the dividing line between, for example, child and adult, is bound to be somewhat arbitrary and the cases barely on either side somewhat arbitrarily classified. Thus, it is useful, at least at the start of a discussion of these issues, to consider individuals noncontroversially within the class in question.[10]

Children

Young children certainly are not fully rational and autonomous (to the degree of normal competent adults) and thus clearly cannot be held to enjoy the basic Autonomy Right of having their destinies determined by their own choices and decisions. However, there is one very good reason for often *treating* them *as if* they had such a right. Treating children as though they are autonomous aids them in *developing* into genuinely autonomous persons; treating them as responsible persons aids them in *becoming* responsible persons. Such "fictions" have a way becoming self-fulfilling prophecies. We shall, of course, want to place limits on such treatment in the interest of the child and pick our instances of fictional autonomy ascription wisely. For example, if a small child decides that he does not want to take the penicillin prescribed for a streptococcus infection in his throat, this would be a bad case in which to let his decision prevail in the interest of developing future autonomy.

Of course, as a normal child grows older, he will acquire greater and greater autonomy. Rights to self-determination can be correspondingly extended, paternalism being phased out gradually. But do not children sometimes, particularly when very young, have a *right* to paternalism – action in their interest either to prevent certain kinds of harms (such as child abuse, subjection to parental superstition with respect to proper medical care) or to provide them with certain benefits (such as a proper education) necessary for having or developing a satisfying life? Here I would answer *yes* – children do have such rights to paternalistic intervention on their behalf. However, the younger and less autonomous they are, the greater the need for intervention – it is *paternalism* after all. What this shows is that the rights here cannot be Autonomy Rights but must be understood as Social Contract Rights. Rational agents in the original position, knowing that they could be born into the class of children and knowing that – even as adults – they will care deeply about any children they may have, will surely want to establish certain minimum floors for children – certain levels of security and education below which no child will be allowed to fall. Thus, paradoxical as it may initially seem, I should argue that there are rights to be treated paternalistically. The paradox disappears, of course, if we remember that we are speaking of rights – not as ways of respecting autonomy – but as legal guarantees society should provide, perhaps as ways of allowing autonomy to develop in those cases where it is not yet present.

Retarded Persons

Autonomy considerations are even less relevant here than they were with respect to children. Severely retarded persons normally have little chance of ever becoming autonomous, and so the "fictional" argument given for sometimes treating nonautonomous children as if they were autonomous will not apply here – at least not nearly to the same extent. Here we are dealing with a class of persons who will never be in a position where it could reasonably be claimed that their destinies ought to be determined by their own choices and decisions; so, if there are rights here, they will for the most part be rights to a certain kind of paternalistic protection.

Are there such rights? Again I should be inclined to answer *yes*, and again the answer is based on social contract grounds. Anyone could become mentally defective or have a mentally defective child (about whom one might care very deeply), and thus rational persons presumably would want to provide a certain minimum floor below which such people would not be allowed to fall. Here what will be relevant is to guarantee a certain level of security and a certain level, not of education, but of training. At the very least (if we have any sense of the good)[11] we shall want to protect them against suffering, including any kind of mental suffering of which they may be capable. The right to be protected against such suffering is like the right to security and training in that it is clearly a Social Contract Right and not an Autonomy Right. In other words, it is simply that which any decent society would guarantee by law to its disadvantaged members.[12]

V. CONCLUSION

At this point, it would be useful to have a tidy *list* of exactly those rights which will be possessed by children and the retarded. I will plead my philosophical vocation as an excuse for not providing such a list, however, since it could be compiled only if we leave the world of abstract thought and begin to gather actual empirical information (which I do not possess) on the exact characteristics of children and retarded persons – what they are in fact capable of at various stages. I have been concerned simply to provide a moral framework in which discussions of their rights can be clarified; and I hope that others, it they accept the framework, will be able to fill in the practical and factual information necessary for its actual employment. At this point the philosopher must give way to the lawyer, the behavioral scientist, and the physician.[13]

I would, however, close with one caution to those who may be inclined to draw up lists of rights for children, the retarded, or any other disadvantaged people. The caution is simply this: Do not give into the romantic and knee-jerk liberal temptation to multiply rights beyond necessity. Autonomy Rights may seem restrictive, but Social Contract Rights are restrictive also – they are, remember, concerned with *minimum floors.* The more one attempts to go above the minimum to assert something more substantial as a matter of right, the more one runs the risk of developing a list of "rights" which are either so economically expensive to support or so morally intrusive to enforce that no sane society will in fact support or enforce them.[14] They will be paid at most a kind of lip service in manifestos – something which cheapens the concept of a right, makes people cynical about rights, and deprives them of their moral force. Social Contract Rights require, not simply noninterference, but *positive* steps to aid. The positive step taken in most complex modern societies, of course, is taxation. If, for example, we are going to have programs to benefit the retarded, someone is going to have to *pay* for such programs. If they are minimum floor programs resting on Social Contract Rights, then taxation may be the appropriate mechanism for funding. If they go beyond this, however, private funding seems in order. When we are under the spell of some good cause, and when we see that the chances of private funding of such a cause are unlikely, we are tempted to manufacture even more "rights" so that we can make our cause eligible for public funding through taxation. But the money available in society is not manna from heaven, to use Robert Nozick's metaphor, failing on the ground to be used for whatever good purpose we see fit. It is *earned* by people and is something over which they may plausibly claim to have a right.[15] Rational persons in the original position would want to protect themselves and their families against certain disasters, and this is why they would agree to minimum floors. Such persons also, however, would surely want to control at least a substantial portion of their earnings to spend as they see fit, and this is why they would be reluctant to agree to anything above minimum floors. Thus, crusades to go above minimum floors as a matter of *right* tend to themselves run the risk of inviting violations of other important rights. If such violations are also economically expensive, they will be opposed on both practical and moral grounds. The final victim of this sort of tension will simply be the important concept of a right itself.

Arizona State University
Tempe, Arizona

NOTES

[1] An earlier version of this essay was originally delivered as a lecture at the 'Conference on the Rights of the Unborn, Children and Retarded Persons', University of Rochester Medical Center, Nov. 19, 1976. It was also presented at the Mountain-Plains Philosophy Conference on 'Human Rights', University of Wyoming, April 1, 1977. I am very grateful for the valuable discussions the essay received at both of these conferences, for these discussions forced me to see important changes that had to be made in the essay.
[2] According to Mill, society's mechanisms of defense are the *law* and the social pressure of *opinion*. In order to simplify matters, I shall concern myself only with the former.
[3] As John Rawls has shown, this theory could be used to justify such morally heinous practices as slavery. See [13], pp. 158–159.
[4] Even if a utilitarian does rule out such sacrifice, he will do so for the *wrong reason*. As Rawls has noted, slavery cannot be wrong because the benefits derived by the slaveholder can never outweigh the burdens imposed on the slaves. This kind of argument fails to appreciate that any benefits derived by the slaveholder are *immoral to count* because they are benefits which are unjustly acquired by the exploitation of human beings and to which he thus has no *right*. Analogously, we do not wait before morally condemning a rape in order to make sure that the pleasure derived by the rapist did not outweigh the pain experienced by the victim.
[5] What follows is not an attempt to give an accurate depiction of the exact details and structure of Rawls' argument in *A Theory of Justice*. As any student of Rawls will

immediately realize, I am collapsing together certain elements in his theory (for instance, the distinction between the stage of the "original position" and the stage of the "constitutional convention") and am skimming over details (such as the distinction between "self-interested" and "mutually disinterested" motivation) which Rawls regards as theoretically important. I am also ignoring some serious problems faced by the theory. While such matters are of theoretical interest, a careful exploration of them is, in my judgment, simply not necessary for present purposes. It would be, indeed, merely a distraction in the present context. Thus the reader should perhaps view the following simply as a set of variations on a Rawlsian theme. The notion of a *minimum floor* which I shall employ is drawn from R. M. Hare's discussion of "insurance strategies" in [4].

One important point of detail, however, should perhaps be stressed. In suggesting that Rawls' theory can be used to provide a derivative analysis of at least certain rights as guarantees that society should enforce, no suggestion is being made that autonomy is unimportant in this theory. On the contrary, the contractors in Rawls' social contract theory are thought of as autonomous in something like Kant's sense; and what these contractors agree to protect or legally guarantee first is a set of certain basic liberties.

[6] See [2]. This is an anthology of critical essays on *A Theory of Justice* and contains a bibliography of other critical essays on Rawls' theory. Two book length attacks on Rawls' theory are [1] and [16]. Though serious and fundamental problems have been located in Rawls' theory, no one has yet, in my judgment, done to Rawls' theory what Rawls has done to the theory of utilitarianism – Rawls has not simply shown its weak points, but also *replaced it* with a theory that is clearly superior. Rawls' theory is certainly not only game in town, but it still seems to me the best.

[7] For Rawls, basic social choice is choice under nearly total uncertainty.

[8] In real, as opposed to hypothetical cases, it is going to be difficult to claim that autonomy is not involved at all. One's status as a rational agent, a maker of free choices and decisions, is going to be impaired by severe hardships.

[9] Whatever the rational solution will be, it will *not* be the highly counter-intuitive one sometimes suggested by Kant – namely, that all autonomy rights (grounding perfect duties) will, no matter how trivial, take precedence over any other moral consideration (perhaps grounding an imperfect duty), no matter how serious. To borrow an example from Joshua Rabinowitz: No morally decent society is going to forbid that I simply take Jones' rope to throw to a drowning person if Jones, who (let us grant) has a Nozickian entitlement to the rope, refuses to let me have it or says I may have it only if I pay him several thousand dollars for it. It is a Kantian perfect duty that I do not take Jones' rope and "only" an imperfect duty that I attempt to save the drowning person. What response can we make to this point except "so what?"

There is insight in Kant's attempted ordering, however, and I think it is this: That the *most important* of the perfect duty requirements resting on the *most important* rights (those which protect autonomy in a serious sense) can be said to take precedence over even the most important of the imperfect duty requirements. Killing Jones, spoiling his health, or taking his liberty away for a substantial amount of time, for example, do not seem acceptable steps to take even to save a drowning person. Taking his rope, however, does not seem to be in the same ballpark with these intrusions. Unless Jones is really weird, the temporary loss of his rope will not interfere with his autonomy in any serious way – it will not keep him from leading one of those meaningful lives which are, according to Nozick, bound up with the concept of rights (see [12], pp. 50–51).

Some of these Kantian insights are captured in Rawls' notion of the priority of liberty, and his claim that his principle of "most extensive liberty compatible with like liberty for others" is lexically ordered prior to all other moral principles. In saying all this, I am admittedly making some assumptions I cannot presently defend: (1) that autonomy rights claims can be ordered on a scale of moral importance and; (2) that interferences with life, health, and liberty are typically more serious rebukes to autonomy than are threats to money and possessions. I think this ordering will have something to do with what I want to call the "internal" relation that one's very self (the person that one is, the meaning in one's life) stands in to one's life, liberty, and bodily integrity, as contrasted with the "external" relation that one stands in with respect to money and tangible things. These are themes I hope to be able to develop at some length in a future paper: to try to make sense of the "internal-external" metaphor (now admittedly just a vague intuition), establish its moral significance, and put it to use. For example, I will attempt to show where philosophers like Nozick go wrong in identifying taxation with forced labor.

[10] A full treatment of these issues would obviously require finer distinctions – with rights graded accordingly – than can be captured by the vague terms 'very young' and 'more then minimally retarded'. Here I am simply concerned to give a general idea of how, in my judgment, discussion of these issues should proceed. For a more detailed treatment, see [10].

[11] I believe that rational agents, as parties in the original position, can have a less selfish conception of the good than that which seems to be stipulated by Rawls. The primary good which Rawls allows them to value is good to *themselves* or *their immediate families* – it is *their* liberty, security, and self-respect which they value and want to secure. The motive for this restriction seems to be the laudable one of wanting to avoid begging any substantive questions in the theory by smuggling controversial moral claims into the model of rational choice. But are there not noncontroversial moral views, views which constitute the "facts" on which morality is based or which are the presuppositions for its very intelligibility? Surely, for example, it is noncontroversial that *suffering is bad* and, within moral limits, is to be opposed. Another way of stating this is that the *interests* of any creature have a *prime facie* claim to satisfaction – something which rational agents who understand what morality is, can surely accept (See [3]. p. 49).

This expanded notion of primary good will have two interesting results. With respect to borderline cases it will, unlike a theory totally grounded in self-interest, make it possible to see the social contract rights extended in these cases as at least in part extended for *their* sakes, and not merely our own. The alternative, as Feinberg notes, is highly counterintuitive. Second, even animals may gain entry into the world of rights given an expanded conception of the good. Rawls himself might want to agree with much of this, for his conception of the good might not be as selfish or individualistic as I have suggested, though it is extremely common for his critics to interpret him in this way. See [14].

[12] Parties in the original position may know that people frequently value objects, such as works of art, and thus may want legal protections for these objects. Does this mean that a stone (if made into a sculpture) may have rights after all? In answering this question *no*, I should want to make three points: First, the Rawlsian apparatus requires our imagining certain misfortunes that might befall us and attempting to protect ourselves against them. It could befall me that I might become mentally disabled, but it seems to

make no sense at all to suggest that it might befall me that I could become a stone. I cannot imagine myself as an occupant of *that* position. Animals pose interesting problems in this regard (see [11]). Second, given the expanded conception of the good noted above, (see discussion note 1 *supra*), parties in the original position will sometimes act for the sake of sentient creatures in order to protect the interests of those creatures. But a stone (even if a great work of art) does not have either a sake nor does it have any interests. Third, my entire discussion is to be understood as accepting Feinberg's conceptual restrictions on the meaningful use of the concept of a right, and I agree with him that ascription of rights to stones is unintelligible. This is why I used phrases such as "creatures have a right to x if and only if" and "individuals have a right to x if and only if" and *never* the phrase "objects have a right to x." The upshot of this is the following: Laws protecting works of art or any other objects do not confer rights on these objects. If any rights are recognized here, it is simply *our* rights as human beings (or the rights of future generations) to enjoy these objects. Here we are acting for our own sakes. See [5].

[13] After reading the present essay, Professor David Wexler of the University of Arizona College of Law called my attention to [15]. Its author, Alan A. Stone, is a Professor of Law and Psychiatry at Harvard University. His discussion, especially at pp. 15–19, overlaps mine to a certain extent and explores some of the interesting constitutional issues arising in this area.

[14] A reasonable minimum floor will, of course, be in part a function of what any particular society, or the entire world, is economically and technologically capable of at a given time in history.

[15] Granted the insight of this Nozickian claim, a qualification (perhaps overly cynical) must be added: Many people with substantial wealth in society have done nothing remotely resembling *working for* or *earning* that wealth as we normally understand these terms. Nozick sometimes seems to attempt to marshal our Puritan sympathies for the work ethic (and even our Marxist sympathies for the labor theory of value) on behalf of people who, though very wealthy, neither work nor "mix their labor" with anything: "Consider . . . how they grow; they toil not, neither do they spin."

BIBLIOGRAPHY

[1] Barry, B.: 1973, *The Liberal Theory of Justice*, Oxford University Press, Oxford.

[2] Daniels, N.: 1976, *Reading Rawls*, Basic Books, New York.

[3] Feinberg, J.: 1974, 'The Rights of Animals and Unborn Generations', in W. T. Blackstone (ed.), *Philosophy and Environmental Crisis*, University of Georgia Press, Athens, Georgia, pp. 43–68.

[4] Hare, R. M.: 1973, Review of *A theory of Justice* (pt. II), *Philosophical Quarterly* 23, 241–252.

[5] Hubin, D. C.: 1976, 'Justice and Future Generations', *Philosophy and Public Affairs* 6 70–83.

[6] Kant, I.: 1785, *Foundations of the Metaphysics of Morals*.

[7] Kant, I.: 1797, *Metaphysical Elements of Justice*.

[8] Locke, J.: 1690, *Second Treatise of Government*.

[9] Mills, J.: 1957, *Utilitarianism*, Bobbs-Merrill, Indianapolis.

[10] Murphy, J.: 1964, 'Incompetence and Paternalism', *Archiv für Rechts- und Sozialphilosophie* 60, 465–86.

[11] Nagel, T.: 1974, 'What is it Like to be a Bat?', *Philosophical Review* **83**, 435–450.
[12] Nozick, R.: 1974, *Anarchy, State and Utopia*, Basic Books, New York.
[13] Rawls, J.: 1971, *A Theory of Justice*, Harvard University Press, Cambridge, Massachusetts.
[14] Rawls, J.: 1975, 'Fairness to Goodness', *Philosophical Review* **84**, 536–554.
[15] Stone, A.: 1975, *Mental Health and the Law: A System in Transition*, National Institute of Mental Health, Washington, D. C.
[16] Wolff, R. P.: 1977, *Understanding Rawls*, Princeton University Press, Princeton, N. J.

JOSEPH MARGOLIS

APPLYING MORAL THEORY TO THE RETARDED

A reasonable, unengaged Martian, scanning terrestrial disputes about moral worth, entitlement, rights, obligations, and the good life would very quickly perceive that earthly creatures had failed to prove that moral values and moral norms can be straightforwardly discovered or that any universal moral consensus of a more than trivial sort can be expected to be achieved. Beyond this, he would find that large convictions of principle were hopelessly deadlocked against one other. He would be bound to conclude that the pursuit of moral principles and moral criteria was ineliminably partisan, embodying one or another such conviction, not itself demonstrably valid or correct – without, however, condemning moral reflection to the irrational for that reason. For example, he would be bound to notice that there is no earthly way to choose convincingly and decisively between the principle that the interests and goods of a limited number of persons may be sacrificed if that would contribute to a significant increase in the benefits and well-being of the majority (an increase not otherwise accessible) and the principle that human persons must without exception be assured a certain minimal measure of respect (so-called human rights) basic to all other pertinent disputes and precluding such sacrifice. In the idiom of the day, these opposed doctrines are said to be utilitarian and contractarian, respectively. The quarrel between them – and many others of the same sort – seriously affects questions of policy regarding the public treatment of retarded persons. But, being human and not Martian, one has the sinking feeling that that quarrel and others like it are cast in such large terms – in universal terms, in fact – that we are actually deflected by them from attending to manageable issues, or human disputes are, ultimately, conceptual frauds.

In the moral context, universality – or, universalizability – takes two utterly different forms: in one, it is really a consideration of logical consistency neutral to and more fundamental than the moral, a consideration conveniently captured by the principle (often misleadingly characterized as a principle of justice) that "similar things must be similarly judged in similar respects"; and in the other, it is really a consideration of global consensus, an idealized consideration of what every rational person would choose or favor, given all the relevant facts (so-called generalizability).[1] The principle

L. Kopelman and J. C. Moskop (eds.), Ethics and Mental Retardation, 19–35.

connected with the first is logically vacuous but humanly (and morally) important – and incontestable. The principle connected with the second is morally substantive but always inapplicable in an empirical way – and ultimately self-serving beyond a measure of moderate plausibility. No one can favor inconsistency rationally; no one has the slightest idea of what every rational being *would* choose; and no one can show that the sacrifice of the putative or actual goods of a certain limited population for the benefit of the rest (or of a good part of the rest) of mankind cannot but be motivated by a lack of respect for the dignity or equal standing of those sacrificed. There is, therefore, no likelihood that the proper moral treatment of the retarded can be straightforwardly derived from any self-evident global principles of morality. One way or another, the same stalemate has been noted with regard to the vexed questions of abortion, war, capital punishment, the treatment of future generations, and the distribution of scarce goods and resources. The pretense that the resolution of such matters can actually be managed by way of an exclusive and compelling argument is simply a tribute to the sincerity of our various illusions as moral partisans.

Now, this charge of course has been disputed. Certainly, those interested in enlarging the range of positive provisions any society should insure for the retarded have found it strategically helpful – as well as morally convincing – to emphasize the equality of rights of the retarded and normal. For instance, they would be likely to support the Declaration of General and Special Rights of the Mentally Retarded (1969) enunciated by the International League of Societies for the Mentally Handicapped, in Sweden, in 1968:

> The mentally retarded person has the same basic rights as other citizens of the same country and the same age (Article I) [9].

Article I of the Declaration corresponds quite closely to the affirmation of the full citizenship and legal rights of retarded persons made in the HEW Report to the President, on Mental Retardation (1976):

> Retarded people have the same rights, legal and constitutional, as every other United States Citizen, including the rights of due process and equal protection of the laws ([18], p. 58).

But what is normally not appreciated is that the actual sacrifice or reduced standing of the retarded need not, in principle, be incompatible with such formulas of equality; also, that the reason this is so confirms the stalemate and the point of the stalemate already noted.

Consider, for instance, that so-called natural or human rights, those applied exclusively to human persons, are controversial in scope, purely formal in intent, utterly vacuous in detail, and entirely compatible with any positive mode of treatment that is itself self-consistent [11]. This may seem too harsh or too broad a charge, indefensible, or at least unfriendly to the protective intent of the articles mentioned. But it is not, and it is important to realize the nature of the limits of all such pronouncements. It is open to dispute whether natural or human rights extend to all members of the human species or only to those who may be said to have achieved the functional level of persons; and it is open to dispute where, precisely, membership in the species or the functional level of persons obtains. One has only to recall current quarrels about the rights of fetuses to appreciate the inherently problematic nature of the issue. Also, whatever we may take to be humane provisions regarding those actually born of humans, who survive, the automatic extension of the rights in question to the comatose, the profoundly senile, the biologically monstrous or profoundly malformed or retarded (see [2, 6]) is at least open to morally responsible challenge. The articles quoted above appear to have decided the issue – within certain operative limits. And so they have, though they have not done so by way of an objective discovery of the proper norms or of anything that could rightly be called a universal consensus. But more than that, they have settled the issue in a way that leaves the essential question as problematic as before.

Two considerations are relevant. For one, the articles quoted above are not expressions regarding natural or human rights but rather expressions regarding the rights of citizens. Human rights – like those of life, liberty, property, respect of person, and the like – are peculiarly vacant. Their importance, apart from the historical influence of certain contingent pronouncements, lies entirely in their collecting more or less reasonably and fully the fundamental prudential concerns of man that any pertinent moral theory would have to accommodate. They are entirely vacant otherwise, in the obvious sense that their support was never intended to preclude war or death caused by self-defense or capital punishment (life), or imprisonment or obligations and duties imposed without explicit consent (liberty), or taxation or eminent domain (property), or even slavery or imprisonment or certain forms of isolation from society (respect for persons); and no self-consistent provisions of any of these sorts can be shown to be necessarily incompatible with the putative equality or inalienability of the human rights alleged. A more positive way of putting the point is simply to say that the doctrine of human rights can never properly be separated from its historical role in promoting certain quite determinate views about the rights and obligations of the members of particular societies during particular intervals of time. However, promoting the latter (special rights or obligations) does not require reference to the former (a doctrine of human rights), has often been indifferent to or ignorant of disputes about the former, and in a sense cannot possibly fail to accord (or to be able to be made to accord) with whatever may relevantly be taken to be the content of the former. It is really the debatable standing of the latter that commands our proper interest. Nevertheless, there is (moving on to our second consideration) no pertinent set of minimal, moderate, or maximal positive rights or goods or provisions that *can* be specified either as a correct determination of the set of inalienable human rights when applied to a given society or state or as a determination based on the universal consensus of rational beings.

Many reasons may be given why this is so; but essentially, the difficulty in question rests with: (1) the fact that the distribution of the resources of any actual, historically contingent, and finite society among other societies depends on perceived scarcities and conditions of survival; (2) the irrelevance of a universal consensus; (3) the conceptual discrepancy between human and positive rights; and (4) the formal compatibility of any self-consistent sacrifice or reduced status imposed on a portion of any particular society by others of that society, with respect both to alleged human rights *and* the rights of citizens. Obviously, the charge that the equality of the rights of citizens does not effectively impose a limitation (or at least any substantive limitation beyond merely procedural constraints) is both puzzling and bound to be challenged. And yet, quite realistically, both the International League's Declaration and the HEW Report to the President concede the point in the most explicit way. The Declaration goes on, after affirming that "no mentally handicapped person should be deprived of [pertinent] services by reason [merely] of the costs involved" (Article II), to say:

Some mentally retarded persons may be unable, due to the severity of their handicap, to exercise for themselves all of these rights in a meaningful way. For others, modification of some or all of the rights is appropriate. The procedure used for modification must contain proper legal safeguards against every form of abuse, must be based on an evaluation of the social capability of the mentally retarded persons by qualified experts and must be subject to periodic reviews and to the right of appeal to higher authorities (Article VII) [9].

The HEW Report pretty well concurs:

Every mentally retarded person, as every other citizen, is presumed to be competent. Only where there has been a full due process hearing can anyone's rights be restricted on the basis of incompetence. The restriction of rights should be limited to the specific problems which the individual has been shown to be unable to handle and which involve possible harm to himself or others ([18], p. 59).

Clearly, the intent is to permit a gradation regarding competence among the retarded to determine the justifiable restriction of citizen rights to particular individuals – consistently with presumed equality and without explicit limits on the extent of such modification. The language is understandably euphemistic; but there is no sense provided (and none can be defended abstractly) that effective "modification" would never amount to effective denial of rights, to substantial differences in the sharing of goods and resources, and even to severely reduced functioning and imminent death. Letting some of the profoundly retarded die (such as cyclopeans and anencephalics [4]) need not violate either natural or citizen rights and, as with some cases of euthanasia at least, seems possible to defend on humane grounds; but the principal issue facing practical morality concerns choices about how to use the limited facilities of an actual society based on its perception of what benefits it can provide its own population and at what cost. Notice that it is almost never argued that the equality of all with respect to natural or human rights *entails* that the resources of the entire earth be made to provide at least minimal – but equal – benefits for all or even for the moderately or maximally competent of the entire earth. How, otherwise, should we explain the world's apparent complacency regarding the starvation of the people of Bangladesh or the Sahel? Contributions to the lives of those people is almost always viewed as supererogatory, hardly ever as obligatory or as directly derived from the inherent worth of, or the respect due to, their members. In fact, the very conviction that there *are* minimal benefits that should be provided in order to bring the quality of citizen life up to a level of bare acceptability goes hand in hand quite easily with the conviction that (even within a society) the lives of those incapable of appreciating or using such benefits may be sacrificed or reduced in quality in order to insure such a level for the rest. There is obviously no entailment here, but, also, there is no incompatibility.

This consideration of actual problems individual societies face explains what is so profoundly mistaken and misleading – and even dangerous – about John Rawls' contractarian system: a view which, in the opinion of many, provides a particularly powerful basis for the protection of the retarded [13, 19]. Relevant to our present purpose, we may say that Rawls' theory of justice fails to demonstrate: (i) that all rational agents are primarily self-interested, in the sense that they would never, under the veil of ignorance, reject the lexical priority of equal liberties of all or subordinate those liberties to the functional well-being of their particular societies or of terrestrial humanity ([16], pp. 42–43); (ii) that, in actual practice, there can be a clear and absolute (lexical) priority given to the equal distribution of "all primary social goods" or that all tolerated inequalities can be shown to be "to everyone's advantage" (preserving the lexical order, even where the distribution of primary goods is postponed) ([16], pp. 150–151); or (iii) that the putatively just arrangements within any actual society are straightforwardly projectible or reconcilable with those of any other distinct or more inclusive society ([16], pp. 520–523; cf [8]). In a word, Rawls does not make suitable provision for the actual scarcity of the goods and resources of any society, or the inequalities of different societies, or for the impossibility of defending the distribution of goods within actual societies without an informed reference to the resources of others and of the entire earth. Hence, he cannot make entirely plausible the restricted scope of the demands of justice (as he sees the matter), or the willingness internal to any society to tolerate inequalities in place in that society; or the preference of a principle of justice under conditions of ignorance regarding one's own place (or the place of the families or parties one represents) within any given system of resources [10].

Rawls himself recognizes that, under the veil of ignorance, justice cannot resolve the moral problem of saving (resources for future generations). "Since the persons in the original position know", he says, "that they are contemporaries (taking the present time of entry interpretation), they can favor their generation by refusing to make any sacrifices at all for their successors; they simply acknowledge the principle that no one has a duty to save for posterity" ([16], p. 140). But Rawls does not see that if this holds for posterity, it must hold as well for other distinct but contemporary societies at least; or that, if the extension to contemporary societies holds, then a similar inequity within any favored society cannot be precluded either; or, ultimately, that the very justice of forming and maintaining a particular society among others (given the problem of the scarcity of resources at least) cannot be decided under the veil of ignorance and takes precedence over all genuine questions of distributional justice. Once the matter is so construed, claims of universal consensus become extremely implausible, certainly irrelevant and self-serving, and the sacrifice or reduced status of particular portions of humanity becomes relatively easy to defend. The very plurality of competing societies, known to possess different ranges and quantities of resources, confirms the effective permanence of sizable societies treating parts of their own populations unequally – within the scope of sincerely defending such practices as morally required, reasonable, or more or less inescapable.

The application of these distinctions to the issue of the rights, abilities, and perceived worth of retarded persons and, therefore, to the obligations of enlightened societies with respect to them is clear. It is quite fair to say that the intellectual capacities of the retarded need not be isomorphic with their capacities to function as responsible agents [19, 17, 3, 1]; and it is equally fair to insist that competence must be presumed to obtain, and struck down only through due process and in a way restricted to particular incapacities. So seen, the denial, "modification", or reduction of citizen rights need not entail any judgment at all of changes in the "perceived worth" of the retarded. The very debatability of capital punishment assures this much: the loss of citizen rights is, as already remarked, entirely compatible with the inalienable nature of human rights – *a fortiori,* with whatever dignity we wish to accord human beings or human persons.

But although it is not necessary, in reducing or denying citizen rights to the retarded, to alter our conception of the worth of such individuals, it is true that to the extent that the ability to exercise one's putative rights personally is construed as essential to the dignity or intrinsic worth of human persons, the justification for significantly reducing or denying such rights may well be thought to entail a changed perception of the competence and worth of the retarded. Here, of course, the now reasonably standard definition of mental retardation becomes pertinent:

Mental Retardation refers to significantly subaverage general intellectual functioning existing concurrently with deficits in adaptive behavior, and manifested during the developmental period ([6], p. 5).

"Significantly subaverage" signifies, generally, an "IQ more than two standard deviations below the mean for the test [administered] " – for the Stanford-Binet, for instance, about twice sixteen to eighteen points below 100, scaled for 'mild', 'moderate', 'severe', and 'profound' retardation. The scales required are of two sorts, varying somewhat independently of one another (intelligence and adaptive ability, open to a variety of dimensions and variable judgments); and the developmental period is taken to range from birth to the eighteenth birthday. The important thing to realize, with respect to worth, is simply this: *if* worth is directly linked to intrinsic human dignity or entitlement to natural rights or the like, then the presumption of worth cannot as such affect at all the differential treatment and management of the retarded; and *if* worth is first linked to actual rational capacity – and only then to dignity or the like – then justifying the differential treatment of the retarded actually determines the measure of their perceived worth. In either case, perceived worth cannot serve as an independent constraint on the treatment of the retarded, except presumptively and for the sake of procedural justice.

How subversive these distinctions may be made to seem is easily judged by considering very briefly the apparently sympathetic view of retarded persons offered by Jeffrie Murphy. Murphy, more or less in the spirit of Kant and Rawls (and Robert Nozick), distinguishes between "autonomy rights" and "social contract rights" [13]. Autonomy rights, Murphy claims, are "particularly fitting or appropriate to *autonomous, rational persons*"; they depend apparently on a "special feature of persons" which provides the conceptually relevant "ground for a special set of moral claims (natural or human rights) which are appropriate only for such creatures and not for any other animals"; and they are clearly linked (in Murphy's mind) with the capacity of persons "to work out *their own* destinies according to *their own* choices and decisions" ([13], p. 6). Murphy goes on to maintain that claims about autonomy rights are "*morally basic*" because all other claims presuppose them ([13], p. 6); and he construes social contract rights in a "Rawlsian-inspired way" such that "An individual should be understood as having a right to *x* if and only if a law guaranteeing *x* to the individual would be chosen by rational agents in the original position" ([13], p. 8). Most autonomy rights are adopted, he says, as social contract rights; but the latter "are going to be broader in scope than Autonomy Rights . . . because rational agents in the original position – in addition to caring about the protection of their autonomy against interference – are going to want to protect themselves against certain major hardships and sufferings" ([13], p. 9). From what has already been said, however, Murphy clearly fails to appreciate the fact that autonomy rights (natural or human rights) are completely vacuous apart from some historical context in which they are thought to be properly embodied in citizen rights; and he fails to appreciate (as does Rawls as well) that social contract rights cannot merely include additions to presumptive citizen rights embodying natural rights and cannot do so *under the veil of ignorance*.

But more than this, Murphy utterly fails to see that the logical upshot of his own theory of autonomy rights – said, you will remember, to be morally basic – is actually to undermine the perceived dignity and worth of the retarded, graded all the way from the mildly retarded to the profoundly retarded (for whom, at last, perhaps *all* dignity and worth must be denied). This, of course, is not Murphy's intention. In fact, he appeals to the principle of autonomy rights in order to expose the "morally bankrupt" status of utilitarianism; for utilitarianism is said to fail "to pay attention to . . . *autonomy* values": "It does not, as any correct theory of justice would", Murphy claims, "rule out the sacrifice of persons for the general good" ([13], p. 7). But, even without regard to the relative plausibility of contractarian and utilitarian theories, it is profoundly ironic that utilitarianism need not, in principle, ignore (even if only in a trivial sense) the dignity of persons; and that Murphy's account, designed to insure such values, actually provides a foundation for their systematic and reasoned denial. For either Murphy must concede that, on his own thesis, *some* significant range of the retarded lack the required ground for autonomy rights (in which case, they may be disqualified from relevant consideration without even "sacrificing" them for the general good) or they have the required property only in a diminished degree (in which case, they may be "sacrificed" to that extent in the interests of those not similarly diminished). It is not merely that Murphy's theory is rendered "morally bankrupt" on its own principle; it is just that it is difficult to show that *any* reasonably popular theory could be convincingly overthrown in Murphy's way, or that, by his strategy, we could really convincingly insure whatever dignity we wished to accord the retarded.

If the foregoing arguments have any force, then it may well be that the most plausible view regarding the worth of human beings (of members of the species *Homo sapiens*) depends simply on that fact: that we are all, regardless of our condition, stage of development, natural capacities, fortunes and misfortunes of chance, achievements and failures, members of a distinct species – from which, at the present moment at least, all known persons develop (whether or not the set is coextensive with the set of human beings); and that it is reasonable to mark for special regard *sans phrase* all such creatures. The trouble with this seemingly humane and unrestricted attitude is simply that *no* substantive conclusions at all follow from or are strengthened by the entailed principle – except, of course, the important procedural constraint of the presumption of equal competence. That is surely not a negligible gain, but it is extremely modest, much more modest than many of its champions have supposed – and ultimately irrelevant to the distribution of the limited benefits of any actual society. It does not and cannot bind us, for instance, to the absolute priority of the equal distribution of any set of social goods; and it does not and cannot specify what any society must determinately provide in order to exhibit suitable respect for the dignity or worth or natural entitlement of human beings. In short, the moral relevance of the doctrine of natural or human rights, or of Rawlsian or Kantian respect for persons, is insured only by denying that doctrine *any* bearing on *any* determinate distribution of social goods – except, always, in the procedural sense noted. Correspondingly, "sacrificing" or reducing, *for cause,* the presumptive interests or citizen rights of a part of a population inevitably increases the resources that may benefit the rest of the population; and yet, such a commitment cannot be shown to have violated, of necessity, any humane view of the dignity of man.

This, then, places the dispute between utilitarian and contractarian principles in an entirely different light. Both are more than dubious; neither is genuinely operative; and neither can insure the reasoned victory of the kind of humane conviction that characteristically directs the work of those devoted to the well-being of the retarded. It is, in this regard, important to understand the potential misdirection of humane impulses – the likelihood, in fact, that the more fundamental and comprehensive and abstract the alleged principle of action, the more readily it may be made to yield unwanted results, if it can yield any results at all. The truth is that the principles and policies required are of a much more restricted sort – consequently, much more accessible, natural, and (inevitably) incomplete as a moral system. The essential idea is that a *liberal* society – that is, a society: (a) committed to some minimal set of equally guaranteed entitlements of citizen rights for individuals, thought necessary for their development and well-being; and (b) committed to some minimal set of equally guaranteed safeguards against intrusions improperly threatening the smooth working of (a) (cf. [16], pp. 72–73 and [15]) – is also committed to progressively enlarging the provisions of (a) and (b), or at least (b), consistently with security, internal order, actual resources, the protection of minimal rights, and the like [12].

We begin, that is, by assuming a certain moral conviction to be in place, one not implausibly (but, also, not necessarily exclusively) imputed to the life of an actual society, whose policies and commitments we are concerned to review *and* to which we ourselves belong. The nature of the liberal principle is such that commitment to it entails a willingness to maximize whatever provisions are thought to embody its particular benefits. There will be disagreements about which goods and resources are the most fundamental and even, within comparatively generous bounds, what constitutes the range of goods that a society ought to insure and safeguard (relative to its actual resources) and how they ought to be secured. There will also be disputes about the liberal principle itself. But one cannot consistently adopt liberalism in any of its various forms and resist the gradual and regular enlargement of its putative benefits within the liberal tradition – its stability and security and resources permitting. Insistence, therefore, on the presumed competence of the retarded already signifies that the discovery of their reduced competence will, consistently with a given society's capacities, insure fuller rather than fewer provisions with respect to (b) and no reduction at least in entitlements with respect to (a) (even if the retarded cannot personally exercise their citizen rights).

Nevertheless, the way in which we must begin here imposes extreme constraints on the exportability of the principles by which particular policies and commitments are to be reviewed and assessed. But that is hardly a disadvantage as far as real-time commitments are concerned. It imposes genuine constraints on the consistency between belief and behavior of a certain determinate population; more fundamental principles are, as we have seen, probably impossible to prove correct anyway; and imputed principles of the same gauge as liberalism must, to justify any measure of tolerance regarding the pertinence of disagreements, be able to be fairly fitted to the same details of the ongoing life of the society in question. So seen, it is a rather simple and straightforward matter to show that, all things considered, a liberal society is committed to enlarging and safeguarding the conditions under which the retarded – both the mildly and the profoundly retarded – may enjoy a measure of well-being. In fact, this finding follows quite trivially from the nature of liberalism itself. Nevertheless, the maneuver is neither pointless nor empty nor dogmatic. Quarrels will arise regarding the application of the liberal principle; reasonable alternatives to the principle will generate further quarrels; and the principle (and its competitors) will be seen to provide internally for open-ended compromises regarding any of its essential provisions.

The most significant gain, then, rests with the fact that the structure of normally pertinent disputes will have been rather clearly articulated. All or most intuitively compelling convictions regarding the treatment of retarded individuals will remain pretty much in place; but their advocates will be bound to organize their arguments in terms of "middle-sized" principles that cannot be interpreted separately from the actual historical traditions in which they are taken to be embedded. Notoriously, what goes wrong with contractarian and utilitarian principles is the absence of any reliable ground for disqualifying or debating extreme hypothetical cases not reasonably linked to the ongoing history of an actual people. There is, for instance, the example of a rather well-known contemporary utilitarian who, on direct questioning, conceded that if lobotomizing the great majority of mankind and giving their care into the hands of a temperate and paternalistically disposed minority would suitably maximize the pleasure and minimize the pain of all (in comparison with other policies) then there was nothing for it but to try, to that extent, to achieve such an arrangement. What is not usually grasped, however, is that *if* respect for the actual moral traditions of a society is to count in any genuine way, then it is quite impossible to assign abstract moral principles like contractarianism and utilitarianism any independent normative force; the best we can expect is that reflection will show a measure of congruence between actual policy and practice and one or another such principle when independently formulated. The inherent limitations of all such principles can be overtaken only in the context of the actual life of a society, relative to which each may already be shown to be (if not entirely compatible with one another) at least plausible idealizations. Anything else leads to the kind of bafflement and polite fraud we have already sampled.

If one requires a source of authority for the adjustments advocated, then the best analogue is probably to be found in Wittgenstein's account of rules and the forms of life of a particular society. For, on Wittgenstein's view, it is impossible to communicate or to share a mode of social life solely by agreeing to explicit rules or principles or definitions governing such behavior: "there must be agreement not only in definitions but also (queer as this may sound) in judgments" ([20], p. 242; cf. also p. 81). Furthermore, on Wittgenstein's view, there is no formal separation or independence possible between the rules governing a practice and the practice thus governed; rules and practices are not hierarchically ordered in the manner of a formal logical system; there are no completely adequate rules for a natural practice, by applying which one could straight forwardly determine how the practice must go on; and practices provide for improvisation, for which sequentially (and variably) projected rules show only how they may be construed as fair (but not exclusive) extensions of hitherto orderly processes.

This suggests the form of actual moral disputes. It warns us, therefore, against expecting too much from moral arguments in support of such disputatious matters as the public care of the retarded; and it instructs us about what may be counted as a genuine gain. Matters like abortion, euthanasia, suicide, war, capital punishment are, like the treatment of the profoundly retarded, inherently problematic. There simply *is* no self-evident moral principle on which the right resolution of familiar disputes about any of these matters can be straightforwardly captured; and any move to search for such a principle invites the hopeless polarizing of conviction that is already too well known. The only prospect of a moderate consensus lies with the frank recognition that there is no discovery to be made, that on such matters it is entirely to be expected that incompatible views will be championed by men of good will. This is why it is hopeless to attempt to derive categorical obligations about the treatment of the retarded from categorical claims about the objective ground for the rights of the retarded. The paradox that needs to be faced is that obligations perceived as categorical are so perceived only conditionally – only within the ideology, only in accord with the principled commitment, of a particular community. This is why, realistically, one falls back to the undemonstrated but reasonably imputed liberal convictions of a society that admit the presumption of equal competence; or why, otherwise, one obliges the advocates of that presumption to formulate an alternative principle (or an alternative argument congruent with that principle) if they demur in supporting seemingly relevant extensions of safeguards and benefits under provision (b), above.

Here, dialectical skill is nearly all there is to count on. Moral argument is remarkably ephemeral at its best, and it requires a measure of approximate convergence of principle and practice to have any force at all – either logical or effective. This is why it is always local, so to say: even disputes between societies must be construed within a shared tradition within which practices and principles may be taken to be common coin. Otherwise, there can be no exchange, only the pretense of a higher revelation. The upshot is that moral dispute cannot eliminate its inherently persuasive and rhetorical function. It would be disastrous to ignore or deny this impurity, if impurity it be. But on that admission, we are bound to see how provisional successful argument must remain. Probably, the care of the retarded arises most manageably, in the United States, within the liberal tradition. If, for example, it is true that "The gap in intelligence and functioning between the more profoundly mentally retarded and the mildly mentally retarded is greater than the gap between mildly mentally retarded and 'normal' persons" ([5], p. 14; cf. [19]), then it should be relatively easier to defend the extension of safeguards under (b) for the mildly retarded than to defend the maintenance of full citizen rights for the profoundly retarded under (a). But even this much holds only relative to liberal presuppositions; and clearly, those same presuppositions may be challenged even radically, without incongruity, within a society used to theorizing about itself in liberal terms.

So disputes about the rights and care of the retarded are an object lesson in the theory of moral argument itself; and in their turn, conceptual discoveries about such argument provide the most remarkably practical moral advice about how to advance the cause of partisan programs for the treatment of the retarded. Disputes about the retarded center on two distinct foci – the worth of a human life, however diminished its competence; and the limits of the resources of a society, however generous its intentions. Commitments regarding these two matters cannot, in actual practice, be separately decided. The pretense to do so leads to the fraud, for instance, that existing inequalities regarding the distribution of basic goods is actually in the interests of all, particularly of the least advantaged; or that those who frankly reduce or sacrifice the entitlements of some in order to provide a closer approximation of what is minimally required for the rest must have violated the inviolable; or that there is some determinate balance between the equal distribution of certain basic goods and the unequal distribution of others that can be shown to be quite correct under the circumstances. Moral dispute, we must finally acknowledge, is dispute among partisans. The only fair constraint to impose is that, as we are rational (being capable of dispute), we ought to be reasonable as well.

– – –

I should like to add a brief note, here, to Jeffrie Murphy's rejoinder to my paper, which I received long after having completed my own [14]. I must say, in all candor, that Murphy has confused two quite different issues and has misunderstood my objection to his own account. *My* objection to *his* theory was simply that, on internal grounds, Murphy's account of autonomy rights was bound to exclude from protection a significant range of just those retarded persons that he would have wanted the rights doctrine to protect – and that there was no way to insure its success. *That's all. He* links the fate of this matter much too closely with another – immensely important – issue, namely, the tenability of the rights tradition itself. There is even the suspicion, in Murphy's statement, that I might actually favor shortchanging the retarded, perhaps by expressing reservations about the rights doctrine. I am inclined to believe, however, that his undefended intuitions and mine converge on the humanity of providing for the retarded.

I freely admit that I doubt that the rights doctrine is either conceptually strong enough, or conceptually the strongest alternative, for mounting a reasoned defense of the intuited policy; I have now just reaffirmed my belief that Murphy's way won't do at all; and I must further admit that I have no confidence in the kind of broad-gauge theorizing that Murphy had favored in his original paper – which he himself now seems inclined to discount somewhat. I have on many occasions, in fact, tried to demonstrate the inherent weaknesses of the human rights theory, though I have always admitted its historically important role (and do so once again, here).[2] If (speaking facetiously, with Murphy) it is Murphy's position that, on any account I might tender, if I do not support the rights tradition more forcefully, the retarded would not even have a right not to be eaten, then so much the worse for the retarded. My view is simply that Murphy needs a stronger foundation for his intuitions, both because his previous defense was so obviously weak and because the rights doctrine won't help in any case.

I must also say that I am *not* a liberal (or a libertarian, or anything of the sort). I did try, I freely admit, to sketch the implications of a liberal theory – and even cited a fuller attempt of mine along those same lines. But I did so because I took the liberal to be defending a rights doctrine. I'm afraid Murphy has seriously misread me when: (1) he takes me to subscribe to the "liberal principle" I sketch; (2) he concludes that because *that* principle may be too weak "the Margolis principle is in trouble." I don't advance any principles of the sort Murphy favors or contests; I don't subscribe to the liberal principle I mention; and I don't believe that principle is actually "vague" (or, for that matter, weaker than Murphy's own principle), though the occasion of the original paper would not have justified a fuller account unless I had actually subscribed to it. To be perfectly honest, *I don't think there is a principled defense of the humane treatment of the retarded that all rational agents would recognize as correct or conceptually compelling; and I don't think it needs one.* Murphy's search for such a principle (I was suggesting) shows only the peculiar vulnerability of risking our humane policies on such a strategy. My own view was simply that recognizing membership in *Homo sapiens* – just that – probably provided the most reliable basis for collecting the humane impulses of human beings toward one another; but that *that was not a principle* (in Murphy's sense) and, in particular, that no substantive conclusions followed from it unless one counts the (tautological) presumption of equality or equal competence *qua* members of the species. I confess I still don't see *what* could possibly provide a stronger foundation for humanizing disputes about the retarded – without, of course, eliminating genuine differences in conviction. But that's just why I think the best we can expect is that partisan disputants may become reasonable partisans.

I must leave the question of the treatment of non-human animals for another occasion. I can't see how the special problems regarding the retarded are illuminated by blurring the distinction between retarded humans and sentient non-humans. I have always believed that that maneuver was a conceptual disaster. I see no reason for thinking that resisting it entails favoring the inhumane treatment of animals, nor do I see any reason for thinking that the differences between the human species and other species are not reasonably plain. The pertinent differences as far as moral reflection about the retarded and non-human animals is concerned are simply: (a) that such reflection is possible in the one species and not in the other; and (b) that such reflection is, in the one case, species-reflexive, and, in the other, not. I have never heard a single plausible argument for supposing that (a) and (b) were not ineliminably pertinent to substantive differences in reflecting on the treatment of humans and animals – regardless of the values one finally championed. I am bound to say, also, that I can see absolutely no force in Murphy's talk of "mandatory" rights except as an extremely tendentious recognition of the species difference between humans and other animals.

Temple University
Philadelphia, Pennsylvania

NOTES

[1] The distinction between universalizability as consistency and as generalizability over the whole of mankind is posed but completely muddled by R. M. Hare [7].

[2] Most recently, in 'The Dialectics of Human Rights', presented (in a symposium including Hugo Bedau and David A. J. Richards) at the Tenth Interamerican Congress of Philosophy, Tallahassee, Florida, Fall 1981.

BIBLIOGRAPHY

[1] Biklin, D. P. and Mlinarcik, S.: 1978, 'Criminal Justice', in J. Wortis (ed.), *Mental Retardation and Developmental Disabilities; An Annual Review*, Vol. 10, Brunner/Mazel, New York, pp. 172–195.

[2] Cleland, C. C.: 1979, *The Profoundly Mentally Retarded,* Prentice-Hall, Englewood Cliffs, N. J.

[3] Cooke, P. E.: 1973, 'Ethics and Law on Behalf of the Mentally Retarded', *Pediatric Clinics of North America* **20**, 259–268.

[4] Cooke, R. E.: 1977, 'The Right to Survive: Relative or Absolute', in L. M. Kopelman and F. G. Coisman (eds), *The Rights of Children and Retarded Persons,* Rock Printing Co., Rochester, pp. 53–65.

[5] Friedman, P. B.: 1976, *The Rights of Mentally Retarded Persons,* Avon Books, New York.

[6] Grossman, H. J. *et al.* (eds): 1977, *Manual on Terminology and Classification in Mental Retardation,* American Association on Mental Deficiency, Washington, D. C.

[7] Hare, R. M.: 1963, *Freedom and Reason*, Clarendon Press, Oxford.

[8] Hoffman, S.: 1981, *Duties Beyond Borders,* Lecture I, Syracuse University Press, Syracuse.

[9] Leland, H. and Smith, D. E.: 1974, *Mental Retardation: Present and Future Perspectives,* Charles A. Jones, Worthington, Ohio.

[10] Margolis, J.: 1977, 'Political Equality and Political Justice', *Social Research* 44, 308–329.
[11] Margolis, J.: 1978, 'The Rights of Man', *Social Theory and Practice* 4, 423–444.
[12] Margolis, J.: 1981, 'Democracy and the Responsibility to Inform the Public', in N. E. Bowie (ed.), *Ethical Issues in Government,* Temple University Press, Philadelphia, pp. 237–248.
[13] Murphy, J. G.: 1977, 'Rights and Borderline Cases', in L. M. Kopelman and F. G. Coisman (eds.), *The Rights of Children and Retarded Persons,* Rock Printing Co., Rochester, pp. 6–20, reprinted in this volume, pp. 3–17.
[14] Murphy, J. G.: 1983, 'Do the Retarded Have a Right Not to Be Eaten? A Rejoinder to Professor Margolis', in this volume, pp. 43–46.
[15] Nozick, R.: 1974, *Anarchy, State and Utopia,* Basic Books, New York.
[16] Rawls, J.: 1971, *A Theory of Justice,* Harvard University Press, Cambridge.
[17] Simeonsson, R. J.: 1978, 'Social Competence', in J. Wortis (ed.), *Mental Retardation and Developmental Disabilities; An Annual Review,* Vol. 10, Brunner/Mazel, New York, pp. 130–171.
[18] U. S. Department of Health, Education and Welfare: 1976, *Report to the President, Mental Retardation: Century of Decision,* Government Printing Office, Washington, D. C.
[19] Wikler, D.: 1979, 'Paternalism and the Mildly Retarded', *Philosophy and Public Affairs* 8, 377–392.
[20] Wittgenstein, L.: 1953, *Philosophical Investigations* (trans. G. E. M. Anscombe), Macmillan, New York.

H. TRISTRAM ENGELHARDT, JR.

JOSEPH MARGOLIS, JOHN RAWLS, AND THE MENTALLY RETARDED

In his paper [6], Joseph Margolis criticizes the attempt to use universal moral principles to understand the moral status of the mentally retarded. In doing so, Margolis has made an important contribution to our appreciation of the difficulties of moral reasoning. A candid appraisal of the capacities of ethics to deliver generally successful moral arguments leads to the conclusion that they are extremely limited. Moreover, when such arguments are delivered, they often support positions we may not have wished in the beginning to embrace. Also, the more one looks, the more complicated things are shown to be. Intellectual honesty will require acknowledging the weaknesses, accepting the inescapable conclusions, and living with the complexities, no matter how painful these may be. It would be much simpler if one could straightforwardly harness a major intellectual account, such as that of John Rawls, to the service of the establishment of the rights of the mentally retarded [8].

It is also worth underscoring why one engages in ethical reasoning. As long as one is not pursuing it solely as a form of rhetoric to convince others or oneself of some cherished opinion, it properly functions as an enterprise of intellectual search and clarification. One engages in it not in order to prove a point, but in order to determine what points can be proven and how. It can be disconcerting when a candid examination does not deliver a straightforward and successful moral account supporting one's initial moral intuitions. And, if Margolis is correct, then ideal-observer theories and ideal-contractor theories do not provide an absolute grounding for moral goals.

In this paper I will in part defend Margolis's Martian, as well as extend Margolis's critique of Rawls. In both cases, I will be elaborating points that are in great measure compatible with the main thrust of Margolis's arguments.

To begin with, the Martian can offer us more than may be clear from Margolis's paper. The viewpoint of the Martian can be employed to remind us of what it means to engage in ethical reasoning. Though the Martian, given Margolis's sketch, appears to share few historical, emotional, and moral interests with us, still, insofar as it is concerned with resolving disputes about the probity of particular lines of conduct without recourse to force, the Martian has implicitly presumed the bare bones of the moral community.

L. Kopelman and J. C. Moskop (eds.), Ethics and Mental Retardation, 37–42.

This is to say, there are two ways in which disputes can be resolved –through force or by appeal peaceably to grounds for common agreement. In short, the Martian can be a member of Immanuel Kant's *mundus intelligibilis.*[1] This view of morality is, I think, best captured in Robert Nozick's view of freedom as a side constraint on action [7]. The Martian can remind us that the minimal condition for morality is an interest in resolving disputes not through force, but on the basis of reasons or inducements. The moral community therefore presupposes mutual respect based on freedom as a side constraint.[2]

As Margolis indicates, this will not deliver much. It will tell one that shooting unconsenting innocent persons for sport is incompatible with this barest notion of morality. It will establish certain obligations to forebearance. It will establish no rights to beneficence. Moreover, the rights it does establish are only borne by moral agents, entities that can understand reasons and freely act upon them according to some moral sense. Though this tells us little of what moral communities are like, it does tell us that particular moral communities will presume the free cooperation of their members. This will establish *strong* rights to procedural fairness for the mentally retarded who are mentally competent, which *most* indeed are, in many areas of life. In short, hard-core moral arguments, those that turn on the very notion of morality itself, can establish rights to procedural justice.

However, as Margolis's criticisms of John Rawls's *Theory of Justice* show, very little more can be delivered *sub specie aeternitatis.* It is useful to put the problem as generally as possible in order to show why any hypothetical choice theory, whether framed in terms of an ideal observer or of rational contractors, must fail. The failure lies in their necessary presumption of concrete moral inclinations or judgments on the part of an ideal observer or rational contractors. That is, in order to decide that some choices are better or worse than others, the ideal observer or rational contractor must have decided that certain social goods are primary and have ordered them in a way that allows moral judgment of actual choices. An inspection of rationality itself will not provide such a concrete moral sense. Indeed, since concrete moral choices require a particular moral sense, the question will be what moral sense to choose in order to justify the choices of the hypothetical contractors, or impartial observer. The regress will be infinite. Thus, hypothetical choice theories function only insofar as their authors project their own moral prejudices onto their ideal observer or rational contractors. John Rawls, for example, could be understood as having reconstructed the moral universe of a liberal member of the Cambridge, Massachusetts, community. One sees this if one pushes somewhat radical considerations. For instance, is it irrational to rank liberty other than first among the primary social goods? What if a loss of liberty greatly increases the goods and pleasures that the other primary social goods insure? Further, is it irrational to put oneself, as a member of the original position, at risk of being in a least advantaged class that would not be benefited by the system of distributing primary social goods, if the risk is small and the goods to be achieved considerable?

Consider a quasi-Rawlsian reconstruction of the society of the BaMbuti, the forest pygmies of Zaire. These forest people live in small bands or villages containing twenty or so families. They are a peaceful people with a deep reverence for nature, who live lives of a hunter-gatherer community [9, 10]. They do not appear to adopt a Rawlsian savings principle. As a result, children born with serious handicaps will likely die. May the BaMbuti morally choose such a life rather than joining the modern society of Zaire? Insofar as one sees their choice as being reasonable, one decides that there are certain goods that can be worth enough to take the risk of being born a member of the least advangaged class of BaMbuti. Or, to put the point more generally, one comes to see that Rawls's *Theory of Justice* has reconstructed a particular notion of risk-taking, with a particular sense of which goods or goals are more important.

The point of these reflections is that the moral life is more created than discovered. There is an element of morality that Martians or rational contractors can discover, which secures fair procedure for moral agents. The concrete structures of a community that give it its particular moral character are those fashioned by that community through history. It is such structures, for example, that secure the position of infants and of the severely and profoundly mentally retarded. The picture is thus complex. There will not be one sense of 'worth' or of 'person', for the meanings of these terms will be determined by the goals that a particular society embraces; they should, however, all include a notion of a moral agent respected through rules of procedural fairness. To appreciate how senses of worth are fashioned in actual societies, consider Aristotle's argument that justice is treating equals equally and unequals unequally.[3] It is only within a particular society, and in terms of its moral goals, that one will be able to determine who is equal, in what circumstances, and what will count as equal treatment. In principle, the only absolute will be the obligation to forbear from unconsented force against the innocent, for this would violate the very sense of the moral community as a community not based on force. However, even this sense of equal treatment must be understood in a particular context in which one will be able to discover what counts as consent, force, and innocence.

If the concrete fabric of the moral life is fashioned through time, how then does one decide which goals to pursue, which moral world to fashion (to borrow Laurence McCullough's metaphor)? Margolis argues that if we embrace the liberal moral viewpoint then we should explore its nature and consequences. This exploration would appear to include choosing among various interpretations of the liberal moral viewpoint. Whether that is true or not, it is here that enterprises such as Rawls's *Theory of Justice* or Bruce Ackerman's *Social Justice in the Liberal State* [1] have their place. They can aid us in seeing the consequences of taking particular moral points of view, and thus endorsing a particular moral sense. But, having clarified the alternatives, how does one then proceed? One will at least, as Margolis suggests, be bound by procedural constraints of fairness. But which of the various possible societies that fulfill the procedural requirements, and which one could achieve, ought one to achieve?

In pursuing an answer to this question, one will not be able to presume that the concrete fabric of the good moral life can be discovered through some intellectual device, given the argument in Margolis's paper. It would appear that at best only abstract lineaments are so discoverable. The concrete moral life is more like the choice of a patron saint [2]. Let us imagine a devout child reflecting on a choice of saints after which to model his life. Let us also imagine that in the end his choice comes down to two: Saint Francis of Assisi, known for his gentle kindness and love of nature, and the patron saint of Norway, Olaf Haraldson, who died in battle. Let us for the sake of argument also grant that Saint Olaf only engaged in just wars and that he never killed with his battleaxe, Hel, anyone undeserving of such treatment. How ought the child to choose? His choice is between two starkly different sets of virtues, which, as all virtues, border on their own special vices. There would appear to be no way to provide general arguments for a definitive intellectual choice. The choice will depend upon the character of the boy and the ways in which the stories of these heroes are told to him. That is, it will not be possible to justify one choice over another in a way that would, in general, resolve conflicts of this sort. The individual who chooses will be able to offer what he or she takes to be good reasons for the choice. However, they will not succeed in being reasons that ought to be generally acknowledged as persuasive outside of a particular moral tradition. The convincing nature of such choices depends upon a particular ranking of goods or values, and such is dependent upon a particular moral sense.

These considerations begin to suggest where our abstract moral reflections bear upon the status of the mentally retarded. Abstract moral arguments can ground modes of fair procedure in the treatment of most mentally retarded. However, they will not establish a concrete moral viewpoint concerning them. That can be done only by showing the moral goods achieved by loving and being loved by mentally retarded persons, by showing how they play mutually morally enriching roles in families, and by showing how they can hold roles of significance as members of a community. As one instructs through morally enlightening stories concerning saints and heroes, stories concerning the mentally retarded instruct as well. One of the important recent achievements with regard to the mentally retarded has been the character of their public visibility showing them in roles that disclose their moral abilities and worth. The Special Olympics are an excellent example. They not only reveal the capacity of the mentally retarded for participation and pleasure in social activities; the Special Olympics also instruct us concerning the possibilities of understanding, appreciating, and endeavoring with the retarded.

Such enterprises disclose moral views about equality, compassion, and attention to the defenseless. As Joseph Margolis argues, moral stories incline persons to the further development of particular moral goods. I have in mind here Margolis's account of the "liberal story" and its commitment to insuring greater guaranteed entitlements that support the development and well-being of all citizens, including the mentally retarded. To be committed to that society is not just to wish to establish certain practices, but to have a moral character of a certain sort. It is an attempt to have certain virtues thrive.

In proferring this account of Margolis's essay, I render his remarks Hegelian and place them in the idiom of some of Stanley Hauerwas's work.[4] They are Hegelian in criticizing attempts such as Kant's to discover ahistorically the fabric of the moral life [3]. One might think, for example, of G. W. F. Hegel's criticism of Immanuel Kant in Section 135 of the *Philosophy of Right,* where he shows Kant's account to be doomed to abstractness. The concrete moral life, as Hegel argued, is to be understood only within the history of an actual society. (I take Hauerwas's idiom of moral stories to be a splendid gloss on this point, at least if one allows for a general grammar for moral stories, i.e., the abstract account. This point is not conceded by Hauerwas.) One comes to understand communities and the moral bias they support through acquaintance with the virtues such communities support. It is through the stories of heroes and saints that it is possible to see, for example, what it would be like to be a virtuous Norwegian warrior of the

11th century. One comes to understand the good of having that character.

The same applies to us with regard to fashioning a community that cares for the mentally retarded through a system of general entitlements. We must choose among alterative virtues and vices.

Baylor College of Medicine
Houston, Texas

NOTES

[1] "It follows incontestably that every rational being must be able to regard himself as an end in himself with reference to all laws to which he may be subject, whatever they may be, and thus as giving universal laws. For it is just the fitness of his maxims to a universal legislation that indicates that he is an end in himself. It also follows that his dignity (his prerogative) over all merely natural beings entails that he must take his maxims from the point of view which regards himself, and hence also every other rational being, as legislative. (The rational beings are, on this account, called persons.). In this way, a world of rational beings (*mundus intelligibilis*) is possible as a realm of ends, because of the legislation belonging to all persons as members" ([5], p. 56).

[2] For my arguments on this point, please see [2].

[3] *See* Aristotle, *Nicomachean Ethics*, iv, 3. He presumes not all persons are equal.

[4] A student of Hauerwas will notice that I have placed his arguments that the moral life is only to be understood within the narrative of particular communities within an Hegelian context, which I take to be their proper place [4].

BIBLIOGRAPHY

[1] Ackerman, B. A.: 1980, *Social Justice in the Liberal State,* Yale University Press, New Haven, Connecticut.

[2] Engelhardt, H. T., Jr.: 1980, 'Personal Health Care or Preventive Care: Distributing Scarce Medical Resources', *Soundings* 63: 3, 234–256.

[3] Engelhardt, H. T., Jr.: 1980, 'Tractatus Artis Bene Moriendi Vivendique: Choosing Styles of Dying and Living', in *Frontiers in Medical Ethics: Applications in a Medical Setting*, Virginia Abernethy (ed.), Ballinger Publishing Company, Cambridge, Massachusetts, pp. 9–26.

[4] Hauerwas, S.: 1981, *A Community of Character*, University of Notre Dame Press, Notre Dame, Indiana.

[5] Kant, I., in L. W. Beck, (transl.): 1959, *Foundations of the Metaphysics of Morals*, Akademie Textausgabe IV, 438–439, Bobbs-Merrill Company, Indianapolis, Indiana.

[6] Margolis, J.: 1984, 'Applying Moral Theory to the Retarded', in this volume, pp. 19–35.

[7] Nozick, R.: 1974, *Anarchy, State and Utopia*, Basic Books, New York.

[8] Rawls, J.: 1981, *A Theory of Justice*, Harvard University Press, Cambridge, Massachusetts.

[9] Turnbull, C. M.: 1968, *The Forest People*, Simon and Schuster, New York.

[10] Turnbull, C. M.: 1976, *Wayward Servants: The Two Worlds of the African Pygmies*, Greenwood Press, Westport, Connecticut.

JEFFRIE G. MURPHY

DO THE RETARDED HAVE A RIGHT TO BE EATEN?

A Rejoinder to Joseph Margolis

Is there any reason which may serve to justify our belief that it would be wrong to kill and eat the retarded – no matter how severe a food shortage might become? (By 'reason' I mean 'consideration which rationally justifies' and not 'factor which causally or psychologically explains'.) Professor Margolis, suspicious of the abstract formalism of moral theory and all its high-sounding talk about rights, prefers to think about such issues in terms of a rather vague "liberal principle" – i.e. the principle that, "all things considered, a liberal society is committed to enlarging and safeguarding the conditions under which the retarded . . . may enjoy a measure of well-being" ([1], p. 29). But are we not committed to a similar principle about non-human animals? Do we not feel that, if possible, domestic animals should enjoy a measure of well-being? But do we not also feel that the retarded should enjoy a moral status somewhat higher and more secure than that enjoyed by our pets? If so, then the Margolis principle is in trouble. It can explain our moral concern about the retarded insofar as it overlaps (as it clearly does) our concern about other sentient creatures, but not as it may represent a special and separate concern. It is this special and separate concern which rights talk seeks to capture; and so, given Margolis' obvious special and separate concern for the retarded, he should perhaps not dismiss the rights tradition so quickly. Perhaps we want to say more than "It is unfortunate that we may have to eat the retarded" (as it might be unfortunate if we have to eat Rover in an emergency). Perhaps we want to say that eating the retarded is *not an option,* that the non-eating of the retarded is a *guarantee.* But is this not simply to say that the retarded have a right not to be eaten? [1]

Assuming the importance of the above question (and extending it to other questions about other possible rights for the retarded), I sought in my 'Rights and Borderline Cases' [3] to explore the question of the extent to which debate about such rights could be rational – i.e. I wanted to get beyond our feeling ("*Of course* we may not eat the retarded, abuse them, experiment on them without their consent, etc.") and see if good reasons could be given in defense of that feeling.

Is species membership alone a good reason? Margolis seems to think that

L. Kopelman and J. C. Moskop (eds.), Ethics and Mental Retardation, 43–46.

it is, but he does not say why and I do not see why. (A mere undefended preference for members of one's own species has been aptly called "speciesism" by Peter Singer – linking it in attitude to racism and sexism.) Species membership can easily seem relevant because it may seem clearly linked to something which is relevant. For example: The Christian view maintains that only human beings have immortal souls and all human beings have immortal souls – no matter how retarded or otherwise deficient they are. The doctrine of ensoulment thus links up species membership with something of apparent moral importance – that precious jewel of an immortal soul. To those nervous about grounding fundamental moral distinctions upon dubious theological premisses, a clear task presented itself: to find some secular equivalent of the doctrine of ensoulment, some special unique "something" about human beings which required that they all receive a special moral treatment not enjoyed by any other animal. Kant came up with this: the capacity for autonomous rational choice. It is this, proclaimed Kant, and not some spooky notion of ensoulment, which explains the special moral status of persons – explains why they have *autonomy rights* (to use the language of my 1977 article). People are autonomous rational agents; other animals are not and thus, it is wrong to kill and eat people in a way that it is not wrong to kill and eat other animals. In short, people have the right not to be eaten. But Kant's view, alas, will not protect the retarded, at least the severely retarded, because they appear not to have the capacity for rational choice. May they then be eaten at will? – or subjected to any other instrumental use to which we may want to put them?

It was my desire to answer *no* to this question, coupled with my refusal to lie about the matter (i.e., to pretend or help maintain the fiction that the retarded have more autonomy than they do have), which promoted my development of what I called *social contract rights* in my essay – i.e. guarantees that a decent society would provide to even its non-autonomous members. I have grown unhappy with this analysis in many ways, but I am not persuaded that it is the total washout that Professor Margolis charges it to be. Even in those cases where honesty compels that we admit that the retarded lack what I called autonomy rights, social contract rights still remain to provide them with certain guarantees which we would not provide for nonhuman animals – certainly more guarantees than would be provided by Margolis' nebulous "liberal principle".

Thus, unlike Professor Margolis, I still believe that rights talk has some important work to do in the area of dealing with retarded persons. It is certainly not the whole story here (as it is not the whole story in any domain of moral concern), but it is part of the story – particularly when our concern is with such questions as this: "How much of our tax revenue will be diverted to provide certain guarantees for the retarded?" Some formal notion of rights may seem a bit artificial here, of course, but it is still surely better than vague (and expensive) sentimentality. There is no free lunch, and every cent we spend on the retarded is a cent we are *not* spending on something else – e.g. the gifted. Rights talk forces us to focus on the question, not of what it would be *nice* to do, but of what it is morally *mandatory* to do. The extent to which our resources are limited will, as Professor Margolis notes, be part of the story here; but this will not be the whole story surely. For even if we have unlimited resources, there is still no reason why we must use them to satisfy illegitimate claims. Thus some notion of a legitimate claim (a right) must be developed even in a world of abundance.[2]

My project in the 1977 essay was to distinguish, with respect to the retarded, morally nice but optional treatment from morally mandatory treatment based on rights (rights which distinguish even severely retarded humans from even very smart members of other species). I still think that the project is worthwhile – particularly if the only alternative is Margolis' flabby "liberal principle." I now believe, however, that my essay did not succeed in bringing my worthy project to a successful conclusion;[3] and thus I believe that part of Margolis' quarrel with my essay is well-taken. I no longer have any confidence in the general program of rationalistic moral philosophy – Kantian, Rawlsian, or otherwise. (See my *Evolution, Morality, and the Meaning of Life* [2] for the details of this change of heart on my part.) I have increasingly come to see these great edifices of reason as explaining the obvious in terms of the obscure – i.e. I have more confidence in any of the actual rights claims I want to make than I do in any of the theories I used to employ to support those claims. This does not mean that I have become a moral intuitionist, however, because I place very little epistemological confidence in that of which I feel confident. Thus I am at a bit of a loss currently as to just what philosophical ethics should be like or about; and I am inclined to think the next important book in the subject should bear the title, not 'A Theory of   ', but 'Muddling Through'. I am still convinced, however, that part of the muddling through will have to involve those protections and guarantees we usually call rights.

Arizona State University
Tempe, Arizona

NOTES

[1] I do not use the example of the right not to be eaten merely to be offensive. I want to raise the issue of rights for the retarded in its hardest context – namely, do they really differ in any significant way from those animals which we feel free to kill and eat? Too much well-meaning sentimentality is allowed to pass for thought in discussion of the retarded, and I want to shock my way through this. Interested parties concerned to sell packages of rights for the retarded often stress that, for example, the retarded have worth because they are loving and affectionate and loyal. Those who argue in this way do not see that these arguments, at most, show that the retarded would make good pets. Surely if special rights packages are in order for the retarded, the case for such programs must be more careful and intellectually respectable than this. It is in this spirit that I argue in this hardnosed way.

[2] Suppose that the plight of retarded persons bores me – that I simply do not give a damn about them or their problems. I know that certain programs would make them happy – would "enlarge their well-being" to use Margolis' language – but I know that other programs would do the same for other groups whose problems do not interest me either – e.g. impoverished stamp collectors who lack the resources to buy the one rare stamp that would make the life of a hobbyist complete. No matter how much money I had, I would be quite indignant if the government began a program of taxation in order to take some of my money and give it to the struggling stamp collector. I would regard this as illegitimate – as coerced beneficence. What *reason* or *argument* could you give me if I opposed, in just the same way, tax money being spent on programs for the retarded? To see this question as central is to see rights as central. "It bores me" is regarded as irrelevant when it comes to footing the bill for what are acknowledged to be real rights – e.g., police protection, free speech, etc. But surely it is a permissible (if not very nice) reason for opting out of all kinds of programs which in some general way expand human welfare. Expanding other people's welfare is simply not always my job; if I go beyond what other people's rights require, I am doing something supererogatory. Thus where does assistance for the retarded belong? – in the camp of respecting genuine rights or in the camp of being nice, acting beyond the call of (public or private) duty? This question is morally and politically important, and Margolis' "liberal principle" fails to come to terms with it.

[3] I certainly did not succeed in drawing a sharp distinction between morally mandated treatment for the retarded and morally mandated treatment for nonhuman animals. I am no longer confident that this distinction is drawable, and thus that part of my earlier project may not be worthy after all. I am still revolted at the idea of killing and eating a retarded human; but I believe I am *just as revolted* at the idea of killing and eating a charming and intelligent gorilla – e.g., Koko. What I happen to find subjectively revolting, of course, is just a matter of my own psychological autobiography.

BIBLIOGRAPHY

[1] Margolis, J.: 1983, 'Applying Moral Theory to the Retarded', in this volume, pp. 19–35.

[2] Murphy, J.: 1982, *Evolution, Morality and the Meaning of Life,* Rowman and Littlefield, Totowa, New Jersey.

[3] Murphy, J.: 1977, 'Rights and Borderline Cases', *Arizona Law Review* 19, 228–241; reprinted in this volume, pp. 3–17.

ANTHONY D. WOOZLEY

THE RIGHTS OF THE RETARDED

In this paper I shall not be concerned, at least directly, with the actual legal rights of the retarded. They have been extensively covered in the literature of the subject, especially in recent decades by authors with whose knowledge and expertise it would be presumptuous of me to compete. What legal rights the retarded currently have is, for the most part, a matter of mainly undisputed record. They all have whatever rights the Constitution gives them, with whatever penumbral uncertainty there may be over the reach of such indeterminate rights as due process, substantive due process and equal protection. They all have their respective state rights varying considerably from one state to another, both in respect of their existence and in respect of their protection and implementation by the state. It does not follow from the fact that they are, for the most part, a matter of record that there are no difficulties about reading the record. For example, since the passage in 1975 (with its amendments of 1978) of the Developmentally Disabled Assistance and Bill of Rights Act [2], much has been written on the assumption that the Act meant what it said when it said that developmentally disabled persons have a right to appropriate treatment, services and habilitation . . . which is provided in the setting that is least restrictive of personal liberty.[1] It was assumed that thereafter states had the duty of providing whatever services and treatment were needed in the least restrictive setting, that a retarded person who needed help had the legal right to receive it outside the walls of a state institution, if his needs could be met outside. But, as many now know, the decision of the U. S. Supreme Court on the Pennhurst case in April 1981 settled that the rights in the Act are only manifesto-rights: the wording of the Bill of Rights section of the Act was too vague to impose a contractual duty on a state [5]. So, what all the legal rights of a retarded person are is not definitely clear, nor likely at any time to be so; and on any question about them there may be good reason why you should listen to what a lawyer has to say, if he is a good enough lawyer, but there is no reason why you should listen to a philosopher, however good he may be.

If a philosopher can contribute any thought at all to the topic of the rights of the retarded, (1) it will be about their moral rights, which will include, but need not be confined to, what should be their legal rights – nothing should

L. Kopelman and J. C. Moskop (eds.), Ethics and Mental Retardation, 47–56.

be a legal right, unless either it is needed as one or it will be useful as one; and (2) it will also be about the appropriateness of the concept of rights to the life of the retarded person. I find something uncomfortably worrying about the latter, and I want to try to communicate some of my discomfort. From now on I am going to use 'a right' to mean a moral right, i.e. a legitimate moral claim which one being can have on another being or on others. That there are such rights at all has been denied, even ridiculed, but this is not the place to argue that; and I am going to assume what would be my conclusion from such an argument, that a being can have rights, whether or not they are recognised, conferred or protected by law.

There is a view, of respectable parentage, which has as one of its consequences that a retarded person, at least one who was severely enough retarded to be identified as such, could have no rights against other, normal persons. That would not be to say that the retarded could have no interests, nor that the normal person was at liberty to disregard the retarded's interests, but it would be to say that there is a sense in which the retarded don't really count, and that, although we should be kind to them, anything that we did for them would be a kindness. Put like that it may sound a pretty odious view, and perhaps indeed it is; but, as I have said and shall indicate, it does have a decent parentage; and it has to be confessed that it responds to something, if not something creditable, in us. It has often enough been observed, and in my opinion truly observed, that the U. S. A. is a country of conformists, and of intolerance, which can reach a paranoid intensity, towards deviants. If the deviants are numerous enough to be seen as a threat to conformity, they become the objects of attack – witness the self-satisfied arrogance of the Moral Majority's castigation of everything that it sees as anti-American, witness the way 'liberal' has become a term of abuse, and witness the way that otherwise fairly rational Americans become rabid if you suggest that communists can be quite decent people. If the deviants are neither numerous nor powerful enough, you do what you can to keep them out of sight: you can't do this with blacks, but you can do it with Indians, and you can do it with the mentally handicapped – put them into state institutions. The sight of someone who is for his years noticeably deficient in intellectual and adaptive functions makes us feel uncomfortable. We both express some feeling of guilt and try to assuage it by denying to ourselves that he is really one of us, and by hiding him away if we can. We try not to see the mother leading her son or daughter with Down's syndrome along by the hand; it upsets us that there should be such people.

Well, as many of you know much better than I do, there are such people, plenty of them – and it behoves us to decide how we stand in relation to each other. Is it true that they, or again at least those of them that are sufficiently bad cases for us to be able to tell ourselves that they are really not one of us, have no rights to be treated or regarded in such and such ways? That they have a number of legal rights is undeniable – so also do illegal aliens, and the children of illegal aliens, but do those legal rights have the backing of justice or the backing of charity? In the case of the illegal aliens it is treated as the backing of justice. Being regarded as, for the purposes of the Constitution, persons, they get its protection; it just happens to be the case, as it might not have been, that the U. S. Constitution is consistently formulated throughout in terms, not of U. S. citizens, but of persons within the jurisdiction. Are the retarded – or again the severely enough retarded – persons within the jurisdiction, that is, are they *persons* within the jurisdiction? Clearly it is going to be difficult to deny that they are, for the distinction between the mildly retarded and the slow-witted, the dense and the unintelligent can hardly escape being capricious, and the scale from the mildly to the severely handicapped is continuous. Yet there might be reasons for denying that all the retarded are persons. Some honorable men have thought so.

One such was David Hume, who two hundred and fifty years ago wrote the following:

> Were there a species of creatures intermingled with men, which, though rational, were possessed of such inferior strength, both of body and mind, that they were incapable of all resistance, and could never, under the highest provocation, make us feel the effects of their resentment; the necessary consequence, I think, is that we should be bound by the laws of humanity to give gentle usage to these creatures, but should not, properly speaking, lie under any restraint of justice with regard to them, nor could they possess any right . . . . Our intercourse with them could not be called society, which supposes a degree of equality ([4], 1.3).

There is no reason to suppose that he was there explicitly alluding to the retarded, but none also to think that he would exclude them, for they exactly match his description of his "species of creatures." His reason for denying that they could have rights, or that our conduct towards them could be characterised either as just or as unjust was that they were so far less than equal to us, especially in the capacity to press their interests. Rights and justice for Hume were fundamentally, if indeed not exclusively, a matter of property; property was a matter of contract; and there could be no contract between two parties, one of which was so much weaker that it could exert no power over the other. This attitude is reflected in one of the two prevailing and competing principles to be found running through the history of the law of contract, the principle that the law can be called on to protect a party from rigorous enforcement of a contract, if he is powerless to protect himself – which is an extension into the law of contract of the principle that some, e.g., those below a given age, are incompetent to make a contract at all. *This* principle is used in many states to rule retarded persons incompetent to enter into business contracts, and in some of those states to prohibit them getting married.

A similar view has been advanced by a philosopher of our own day. John Rawls identifies "the features of human beings in virtue of which they are to be treated in accordance with the principles of justice" ([6], p. 504) with "the basis of equality," and, like Hume, he excludes non-human animals from the scope of rights; there should be constraints on our conduct towards them, but they are not the constraints of justice, the regard for rights in animals. But, whereas for Hume the relevant equality was equality of power, for Rawls it is equality in moral personality. A being that has a moral personality, either actually or in capacity, has rights; and to have a moral personality is to have a sense of justice. While Rawls is more cautious than Hume, in that he does not commit himself to the view that a being that does not have a sense of justice cannot be owed the duties of justice, and while he avoids the issue raised by the case of the severely retarded by saying that it "may present a difficulty", at the same time assuming that "the account of equality would not be materially affected" ([6], p. 510), he does in the end say "it does seem that we are not required to give strict justice anyway to creatures lacking this capacity" ([6], p. 512), i.e. the capacity for a sense of justice. What this comes to is (1) that it is possible for us to treat unjustly, and that we have the duty not to, a being which has a sense of justice, i.e. one who can be aware that it is behaving or is being treated unjustly; and (2) he is inclined to say, but just holds himself back from outright saying it, that a being which lacks this sense cannot be treated unjustly – although it can be treated cruelly, callously, heartlessly, etc.

It may seem that, whatever we say about Hume's view, we have to say that Rawls's does not work against the retarded. We may answer that we may not know what to make of the claim that a battery hen or a calf in a veal-unit, or even the family dog, has a sense of justice, but a retarded being does, and so he has a *right* to be treated in such and such ways by those whose lives cross his. The trouble with that answer is that it is simply untrue. A retarded person does not lack a sense of justice by being retarded – unless he is retarded enough; and surely plenty are retarded enough for that. Such a person will be bewildered, distressed at the way people treat him, as being a nothing of some sort, as something to be addressed like a grown-up child, something that can be talked about in his presence as if he was not there – but he does not have the comprehension which you must have to have a sense of justice. A dog can look at you pleadingly, or even perhaps accusingly; but to say that he is pleading for justice, or accusing you of injustice, is to attribute to him a concept which it would be rash to suppose that he has; the same must be true of many of the retarded.

But why should we suppose that, to have a right, a being must understand what it is to have one, and to have it respected or violated? Many beings besides the retarded have legal rights, but lack the capacity, whether actually or potentially, to grasp their entitlement to have those legal rights met – e.g., the too young, the too senile, some of the mentally sick, and all of the terminally comatose. Are we to say that, although they have their legal rights, they do not have the moral right to have their legal rights respected? If we cannot say that of them, we cannot say it of the severely retarded either, for they are in the same position. The question here, right across the board, is what do you have to be to be a holder of rights? And yet that seems not to be a well formulated question. It suggests, first, that a right is some kind of an entity which you either have or do not, like an identity card, or a driver's license, or a credit card – that it is a special kind of possession, which you must be old enough, or bright enough, or something-else enough, to have. An answer of that sort is unpromising, not only because it sets rights up as some kind of metaphysical snarks, the hunting of which will be unending and fruitless, but also because it has us starting on the wrong foot. The question whether a being, which is different enough from the rest of us, who, we do comfortably suppose, have rights, does or does not have rights is less a question about it than it is about us – what our attitude to that being should be.

Fifty years ago "three generations of imbeciles" seemed enough to Justice Holmes and the Supreme Court to justify the compulsory sterilization of women who were unfortunate enough to be institutionalized in, of all states, Virginia [1]. That we are, barely two generations later, appalled at that judgement has little to do with our seeing it as based less on medical evidence than on a fearful prejudice against the visibly different and subnormal; after all, most of us are as ignorant as Holmes and his advisers about the actual facts of the transmission of mental defects. It has much more to do with a widespread change of social attitude. We do not see the sterilizing of women as if it were the spaying of bitches (and maybe in another two generations or so *that* will seem appalling). We do not kid ourselves that it is all right because they will not mind, and because doing a cost-benefit analysis is much easier: the good of preventing them from continuing to "sap the strength of the state" will easily outweigh the evil of the "lesser sacrifices" which they have to make, especially if the sacrifices are "often not felt to be such by those concerned". But who says that those concerned do not feel it, and that so we need not worry? If it really was believed by those on the spot that *they* would not feel it, why were so many of them not given the chance to feel it, but were told that the surgery was an appendectomy?

We may not have come far in fifty years, but we have come far enough to see the mentally handicapped as having feelings and a sense of life, and as entitled as the rest of us to whatever independence, education and freedom they can responsibly handle, and, in the case of those less capable of looking after themselves, protection from the omniscience of some bureaucrats and of some physicians, e.g., those who not only do not have scruples, but cannot see that there is anything to have scruples about, over nontherapeutic experimentation.

So far so good, but so far already, I suspect, some bad. This is going to be difficult to put across, because, like most of us, I lead a sheltered life and do not have enough exposure to the disquieting to be able to handle it confidently. The people we are talking about are a minority, and are kept, whether privately or publicly, a highly inconspicuous minority. Most of us have as little familiarity with them as we do with the inmates of our prisons, whose lives our lives never cross, except, say, when they break out and scare us into helping them run – and to how many of us does that happen? We all know, or know of, a family with a mentally retarded member, but we do not know him/her, and would be totally ill at ease, clumsy and even frightened if left alone with him for half an hour, just as we would be if left alone with a prisoner in his cell for half an hour. Usually those close to a retarded person, say a child victim of Down's syndrome, will care for him very much – brothers and sisters are devoted to him, and his mother ferociously protects him with her love against the well-meant efforts of others, who insist that he would be better off in "a home". But the rest of us are outsiders, and maintain our distance as much by arguing for, and insisting on, the rights of the retarded as by denying that they have them. This is where I get to the bumpy part of my paper, and where I may alienate a sizeable portion of my audience. Almost everything that is written nowadays on the subject is in support of the retarded person, and almost all of it I find repellent, although I know it is intended to be on the side of goodness and light. It is a part of what happens when one focusses attention on rights. It is, therefore, part of a much larger and, if I may say so, especially American problem. Let me introduce this by quoting one sentence, or part of one sentence, from the Report of the Task Panel on Legal and Ethical Issues to the President's Commission on Mental Health in February 1978. It reads: "There is now a revolution of expectations of the mental health consumer . . . ." [7]. That may sound harmless enough, but it worries me, not just because of its barbaric English, but because of the corrupted thought which that barbarism manifests. When a mentally retarded man or woman becomes a mental health consumer, something is going wrong; and one of the things that is wrong is that people who can write like that do not see it.

Here we are brought up against a tension, and the distortions which it introduces, which is a feature of our way of life, and which is going to remain so – unless the bomb comes along to reduce those who may escape its annihilation to a newer (or perhaps older) relationship with each other. For practical purposes, the tension must be regarded as permanent and unavoidable, but the distortions could at least be reduced by acknowledging the tension, and by a change in attitude which would do something to counteract it. The tension is that between, on the one hand, people's needs as the people that they are, and, on the other hand, the organization of provision to meet those needs, of weighing the needs of some against those of others, of the decisions which to prefer to which other – organization which is a necessary part of any even halfway sophisticated and halfway civilized society that has submitted itself to the rule of law. We can moralise – and so much good moralising on these things has already been done that it would not be either illuminating or stimulating for me to do more – on the rights of the retarded, how far their autonomy is to reach, how far protection of them is to go, etc. But to do that is to miss a point which I am not going to claim is *the* point, the only point, but which I do suggest might lead to a better state of affairs if it were treated as if it were.

The tension I have referred to is that between people being the individual people that they are, each with his or her own needs, emotional tendrils, aspirations, ties (or the lack of them) to family, friends and personal heroes, and, on the other hand, the law and its agencies seeing them as members of a class – children of school age, juvenile delinquents, the mentally retarded or whatever. This tension is not peculiar to the individual-and-the-law, it can be seen in the relationship between the individual and many other social and economic institutions, e.g., the insurance companies: traditionally, a woman, on retiring, finds that the contributions which she has over her working years paid into a pension fund have earned for her a lower annuity than a male receives, having made the same contributions. For the actuary she is not the person she is, she is a member of a class which statistically lives longer than the statistical male. This classifying of us is an unavoidably pervasive feature of organized social life, and it breeds, especially in such a litigious society as this, what I find a disturbing preoccupation with rights.

Rights are things that we have as members of a certain class, the class usually being defined by our status or our role. A being does not have rights – period; he has rights as a ___, as a policemen, as a taxpayer, as a resident alien, as a parent, etc. In the same way a non-human animal has (or so many suppose) rights as a sentient being, and a retarded person has rights, some as a human being with certain capacities, others as a human being with certain incapacities. Legislators, and their medical and social advisers, have to decide what legal requirements to impose on the community for the benefit and protection of the retarded, what legal liberties to confer on or confirm for the retarded – all this without knowing, or knowing of, the retarded except as members of a class, viz. the retarded, with as many subdivisions of the class (the severely retarded, the mildly retarded etc.), as is practically and administratively convenient. This has to be; this is the way the world is run, if we have to decide what to do to people, for them and against them, when we are not personally involved with them at all, when they are people whose names are on a list, or whose Social Security numbers are on a computer printout, or who are included in an institution's total of current inmates. This is a way, and a legitimate way, to think about the retarded, but to make it *the* way to think about them skews the whole picture; it is what leads to thinking of them as mental health consumers. It is true of you, I hope, i.e. you who are reading this volume, that you are somewhere above the lowest level of intelligence of those who in our society are allowed to circulate on their own; and is also true – or I should be very surprised if it were not – that among you the limits on intellectual and adaptive function are no more uniform than they are among the retarded. We are different from the retarded, in that our levels of capacity to perform are, given the way we make our social assessments, higher than theirs. What each one of us prizes most in himself/herself and regards most in those others whom he knows well enough to regard at all is being, and being allowed to be, not the *kind* of person that he is, but *the* person that he is. What I am trying to get at is that we care about the people we do care about (and we want to be cared about ourselves), not as people of a certain kind or class, but as the individuals that they are, and in the relationships to ourselves that they are; and it does not affect our caring that we have misread the person or the relationships. There does seem to me to be real danger that in all our thinking about, and solicitude for, the retarded, their needs and their rights, we tend to forget that each of them is somebody, he is not just a member of the category of the handicapped.

I have come across two horror stories recently, which bring out the humanly awful outcome that well intentioned and carefully thought out decisions can have. The first is not about the retarded at all, but it illustrates the point I am trying to make. In 1981 a couple (in England, I think) were not allowed to adopt the child they wanted to adopt, on the ground that their home would not provide a suitable environment for the child's upbringing; and the reason why it was judged unsuitable was that this couple were a notoriously happy and harmonious pair – they never quarrelled or fought: such a home was not the right place to prepare a child for life in the real world. The other story is American, and concerns a girl of 16, with the mind of a 3 year old. On the recommendation of their local school board her parents, with great misgivings and anguish, some years ago placed her in a residential school for exceptional children. But it was a success, and, when the family moved from New Jersey to Virginia four years ago, they transferred their daughter to a similar private residential school in Maryland. The success continues: by both her mother's account and that of the school's director the girl "is secure, she's happy, and she's progressing more than we ever expected her to " [3]. Yet their new school board (Fairfax County), having failed to persuade the parents to transfer her to a daytime public school for the retarded in the county, then sued them to force the transfer. Expense was not at issue: the educational cost to the school board would be much the same in either case. The Fairfax board took its stand solely on the "least restrictive setting" notion, arguing that her present school, being on a 24-hour basis, was too custodial and restrictive, as contrasted with the 6-hour-a-day program provided by the county's own school, and "fails to prepare her for daily life in the outside world." The girl they were talking about is so severely retarded that she "cannot speak, cannot read nor write, and is only now learning to control her bodily functions." They did not dispute that she is happy and making good progress where she is, they did not ask themselves how *she* could hope to cope with daily life in the outside world. They just had a classificatory dogma for education of retarded children in the least restrictive setting; and that was it. Fortunately, thanks to the good sense of the judge, the story had a happy ending. What I find disturbing about that affair has nothing to do with obtuseness or bureaucratic pigheadedness with which some might want to charge the county. It is rather what seeing the retarded as bearers of rights and consumers of services can do, if that becomes the *only* way of seeing them. The rights of the retarded are important; what is more important, and should not be forgotten in the earnestness of providing for them is that there should be in each case, whenever it is humanly possible, somebody who loves her or loves him, and to whom talk of rights must seem strangely inappropriate.

University of Virginia
Charlottesville, Virginia

NOTE

[1] *Pennhurst State School v. Halderman,* 451 US 1 (1981). In a later case (1982) the Court decided that a mentally retarded inmate of Pennhurst State School did, under the due process clause of the Fourteenth Amendment, have constitutionally protected liberty interests *inside* the institution to "resonably safe conditions of confinement, freedom from reasonably bodily restraints, and such minimally adequate training as reasonably may be required by these interests." *Youngberg, Pennhurst State School et al. v. Romeo,* 50 USLW 4681 (US June 18, 1982) [8].

BIBLIOGRAPHY

[1] *Buck v. Bell,* 274 U. S. 200 (1928).
[2] Developmentally Disabled Assistance and Bill of Rights Act: 1976, 42 USC 6010.
[3] Hochstader, L.: 1981, 'Retarded Fairfax Girl Is Caught in a Legal Tug-of-War', *Washington Post,* July 20, 1981, p. B1.
[4] Hume, D.: 1777, *Enquiry Concerning the Principles of Morals.*
[5] *Pennhurst State School v. Halderman,* 451 US 1 (1981).
[6] Rawls, J.: 1971, *A Theory of Justice*, Harvard University Press, Cambridge.
[7] U. S., Report of the Task Force on Legal and Ethical Issues to the President's Commission on Mental Health, 1978, U. S. Printing Office, Washington, D. C., V. C. 1. Reprinted in 20 *Arizona Law Review* 49 (1978).
[8] *Youngberg, Pennhurst State School, et al. v. Romeo,* 50 USLW 4681 (US June 18, 1982).

CORA DIAMOND

RIGHTS, JUSTICE AND THE RETARDED

Is proper treatment for the retarded a matter of justice or of charity? – That is the issue, raised by Professor Woozley [8], on which I want to comment, but I shall start at some distance from anything he discusses. There is an idea which can be found in quite a lot of recent writing about the rights of the retarded: the idea that *we* are finally getting away from the myths and ideologies and irrational fears which distorted our ancestors' ways of thinking about the mentally retarded. This idea is, for example, expressed by Paul Friedman, well known for his legal work on behalf of the retarded, and by Issam Amary, in a book on the rights of the retarded ([3], p. 16; cf. [6]; [1], pp. 3–4). Amary states that traditionally the retarded have been regarded as outcasts of their societies and as individuals who brought shame to their families; and Friedman says that traditionally the mentally retarded have been viewed as subhuman organisms or as menaces to society or as eternal children or as irreversibly diseased persons. The view that you get then in writers like Amary and Friedman is that while there still are people who think in terms of such stereotypes, we have at least made a beginning at getting away from such false views. We have at least begun to recognize retarded people as our fellow human beings; we have begun to recognize their rights, and to modify and reshape our social institutions to accord them their rights.

No doubt some – or many – of our ancestors have had pernicious and false views about the retarded. But if we think of our ancestors as all blinded by prejudice, *we* are being blinded by prejudice about them, and we will not learn what we can from the past and from traditions still alive in other parts of the world.

Let me contrast with the picture we get in writers like Friedman and Amary two descriptions of the treatment of the retarded, one a description of a nineteenth century Russian town, the other of a traditional culture surviving into the present. The first description is a well-known one, from Part I, Book III of *The Brothers Karamazov,* the description of the life of Lizaveta, the idiot [2]; The description is fictional but is meant to be a realistic picture of at any rate many Russian towns. Lizaveta is described as virtually incapable of speech, dwarfed, with a look of blank idiocy and a

L. Kopelman and J. C. Moskop (eds.), Ethics and Mental Retardation, 57–62.

fixed stare, unpleasant despite its meekness. She slept in the dirt in all seasons and wore nothing but a smock. Her mother had long been dead, and her father was a drunkard, who would beat her whenever she went back to him – which, however, she rarely did, "because everyone in the town was willing to look after her as being an idiot, and so specially dear to God." They would give her warm clothes for winter, and she let them put the clothes on her, and then would take them off and leave them somewhere. When her father died, people looked after her even more than before, the religious ones, as she was now an orphan. Liked by everyone, she would walk into strange houses and was never driven away; "everyone was kind to her and gave her something" – but what she was given she would give to the first person she met. She would climb into people's kitchen gardens and sleep there, and in the winter would sleep in the outbuildings of her father's old employers. Much to everyone's horror, she is raped – by a drunken man, on a horrible drunken dare – and becomes pregnant, but is very well looked after during her pregnancy, and a wealthy woman arranges for her to be taken into her home and looked after during the birth. And when the baby is due, they try to keep watch on her. She gets away, however, on the very last day, climbs over a high wall into a garden and is injured while doing so, and dies in childbirth ([2], pp. 95–98).

There are two features of the story to which I want to draw attention. First, the idiot woman is *not* thought of in accordance with any of Paul Friedman's traditional models: Lizaveta is not a subhuman organism, a menace, a child, or irreversibly diseased. She is thought of as specially dear to God, even more so as an orphan. Secondly, she is not locked away out of sight, nor is she left to fend for herself. The view of that traditional community is that the care of the idiot belongs to the community.

As I said, this account is, though fictional, based on Russian reality; and Lizaveta's treatment has indeed got similarities to the way retarded people have been treated traditionally in other parts of the world. Some such similarities come out if we look briefly at a second example, the treatment of the retarded in a traditional culture surviving into the present, that of the Indian subcontinent. My information here comes from a Muslim, a political scientist who has lived and travelled in many parts of India, Pakistan and the Middle East, and whose description was intended to cover both Hindu and Muslim communities (Eqbal Ahmad, personal communication). He made two points about the way retarded people are treated on the subcontinent. First, in comparison with the West, it is very striking that there are absolutely no remedial educational facilities of any sort whatsoever; in *that* respect modern societies offer something better than anything in traditional ones.

Secondly, though, in the traditional societies he knew, retarded children and adults are much better integrated into community and family life than they are in the West. This is partly, but *only* partly, a matter of their being drawn naturally into communal work. Weaving and the making of mats and rugs and the like is important in these communities and involves much repetitive work of a sort retarded members of the family can handle. More important for us to note, though, is the central idea that these communities have about retarded people, an idea my informant came back to several times: this is the idea that the retarded person is a special blessing of God on the family. That central idea is related to a group of others. The retarded child may be viewed as someone whom God has afflicted, has allowed to undergo a sort of punishment, for the mistakes or wrongdoings of others in the family; he is a test of our ability to take in others with compassion, an opportunity for us to acquire merit in the eyes of God by doing good for those who are not fully capable themselves. My friend has a retarded relative, and he told me how his family, in their relationship to this retarded young man, was spoken of by a temporarily hired chauffeur, talking to other servants. Wonderful, such good people, he called the family, blessed by God with a retarded child, and God has rewarded them for their gentle treatment of him by happiness and well-being.

My point in giving these two brief descriptions is *not* two things. I am *not* suggesting that we should regard the treatment of the retarded as better in traditional societies than in ours; I want to make no judgment better or worse. I am also *not* suggesting that our ancestors, all of them or even most of them, regarded retarded people in the ways I have described. I *am* however suggesting that the picture we may have, that, traditionally, retarded people have been regarded in terms of stereotypes that put them down, and expressed a combination of contempt and fear, is – taken as a generalisation – simply itself a prejudice, and a misleading stereotype. I have wanted to bring out that there are present (in some traditional societies anyway) modes of sensitivity to the needs and feelings of the retarded, modes of sensitivity in some ways strikingly different from ours and from which we may have something to learn.

What is most striking about the two cases I have mentioned is what they have in common. Despite their representing such different cultures, nineteenth century Russia on the one hand and India and Pakistan on the other, in both we see use being made of the idea of the retarded person as in a special relation to God, specially dear to God, or a blessing from God, and this idea is in both cases connected with the idea of the welcomeness of the retarded person in the community at large. In our own English-speaking tradition we have in fact a related idea expressed by Wordsworth. The helpless and the dependent (he tells us) cannot be without offense to God cast out of view, put in institutions out of sight. They bring blessing to the community in ways we can only with insensitivity forget [9].

I now want to explain, and I shall have to be too brief about this, how what I have said is related to Professor Woozley's paper. He is critical of much current thinking about the retarded. It too exclusively regards retarded people as bearers of rights and consumers of services, he says. With this I entirely agree. However, he relates his criticism to a wider question, asked earlier in his paper. This is the question whether the moral basis for proper treatment of the retarded lies in justice or charity. He refers to John Rawls, who asks whether the constraints that there ought to be on our conduct towards the retarded are constraints of justice and regard for rights, or on the other hand merely constraints of compassion and humanity ([5], p. 512). The philosopher Jeffrie Murphy also asks: is the decent treatment owed the retarded a matter of charity or of their *rights*? [4]

It is on this issue that I think we have something to learn from traditional societies. But what *is* that issue, exactly? The philosophers whom I have just mentioned put the issue in two different ways. They ask whether the constraints that there should be on our treatment of the retarded are constraints of justice or merely of humanity; and they also ask, as if it were exactly the same question, whether the constraints on our conduct towards the retarded are a matter of respect for their rights or merely a matter of humanity and so on. They identify treating retarded people justly with respecting their rights, and treating them unjustly with disregarding rights.

I want to suggest that we should not tie the notion of justice to that of rights; and that treating the retarded with justice has an importance we can recognise only when we get clear about the difference between asking for justice for someone and demanding that he be given his rights. Justice is the more powerful and important notion, and is weakened if tied too closely to the demand that people be given what they have a right to. We can treat someone unjustly by failing to respect his rights; but also, surely, if we take advantage of his weakness. Taking advantage of those weaker than oneself, taking advantage of those who are especially vulnerable or dependent – this is a despicable kind of thing, a central case of injustice; and it does not need any kind of backup in terms of rights. When someone takes advantage of a vulnerable person, the crucial objection is not that he has not been given his rights. – Traditional societies may link the idea of harming the specially vulnerable to the notion of arousing God's anger, to the notion that he cares specially for the retarded and is thus especially offended by harm done to them, or scorn to them, by such injustice. The society or the family that treats such people badly cannot prosper: the good of community and family are linked to that of its most vulnerable members. Even if only in providing opportunities for other people to act with gentleness and generosity, the vulnerable and dependent contribute to the good of others: not to recognise this, and to regard such people as a mere drain on communal resources is itself a kind of blindness, linked closely to injustice – and also, and importantly, to discourtesy.

I want to make a final comment on where this leaves the question whether proper treatment for the retarded rests on justice or compassion. Let us first be clear that there is a difference between asking for justice for someone and asking on his behalf that he get his rights. Let us also be clear that compassion, charity and humanity must not be confused with the impulses that move the do-gooder or the philanthropist. The essential impulse in charity – in *caritas* – is the impulse to attend as unselfishly as one can to the reality of another person, to perceive his needs and to feel what it is for him to be hurt or bewildered or scorned. Justice worth the name depends on that too. The question whether the constraints on our behaviour towards the retarded are those of justice or of charity is a false question. They are far more closely linked than that question allows.

(I have drawn in these last two paragraphs from ideas expressed by Simone Weil in 'Human Personality' [7]. It does not seem to me that questions about the rights of the retarded can be put into proper perspective without a full discussion of the questions she raises.)

University of Virginia
Charlottesville, Virginia

BIBLIOGRAPHY

[1] Amary, I. B.: 1980, *The Rights of the Mentally Retarded-Developmentally Disabled to Treatment and Education,* Charles C. Thomas, Springfield, Illinois.
[2] Dostoevsky, F.: 1879, *The Brothers Karamazov.*
[3] Friedman, P. R.: 1976, *The Rights of Mentally Retarded Persons,* Avon Books, New York.
[4] Murphy, J. G.: 1978, 'Rights and Borderline Cases', *Arizona Law Review* 19, 228–241, reprinted in this volume, pp. 3–17.

[5] Rawls, J.: 1971, *A Theory of Justice*, Harvard University Press, Cambridge, Massachusetts.
[6] Roos, P.: 1973, 'Basic Facts about Mental Retardation', in B. Ennis and P. R. Friedman (eds.), *Legal Rights of the Mentally Handicapped*, Practising Law Institute, Mental Health Law Project, New York.
[7] Weil, S.: 1962, *Selected Essays*, R. Rees (ed.), Oxford University Press, Oxford.
[8] Woozley, A.: 1984, 'The Rights of the Retarded', in this volume, pp. 47–56.
[9] Wordsworth, W.: 1800, 'The Old Cumberland Beggar', in *Lyrical Ballads*.

SECTION II

RESPECT AND LABELING

LORETTA KOPELMAN

RESPECT AND THE RETARDED: ISSUES OF VALUING AND LABELING

In the nineteenth century, there lived a monk known as Brother William, who was a man of very limited intellectual abilities. Brother William believed he was in constant contact with God. His life was so exemplary in its kindness, compassion and love that he was greatly honored in his community.[1] Brother William sounds as if he was a person who deserved esteem, and whom we should hesitate to call handicapped even though he was, by present criteria, mentally retarded. But language itself shows, despite what we like to think, that (outside of families) we rarely give retarded persons respect in the sense of esteem.

Even to lump together so diverse a group under the locution 'the retarded' shows our attitude to those we think of as failing below the standard of normal intelligence. At one end of the spectrum it includes those who just fail to be honored by the title 'normal', and who, by other standards might be considered normal or able to function perfectly well. (Insofar as the line between normal and below normal is arbitrary, many persons could be "cured" by redefining 'normal'.) At the other end of the spectrum, it includes those who are so profoundly retarded they seem forever incapable of actions, hopes, aims, beliefs or of having a sense of time or of self. Though individuals who are called retarded are extremely varied, the language used to describe them is not. For while the range of non-technical expressions we use to indicate the different kinds of ability and intelligence of those between borderline normal and genius is rich and subtle, the retarded tend, unfairly, to be regarded as one group; or if distinguished, they are described as falling into one of four groups: mildly, moderately, severely, or profoundly retarded.

Language also betrays general social attitudes to retarded persons when we consider how quickly terms intended to describe them are transformed into pejorative expressions. The now obsolete and derogatory terms 'moron', 'imbecile' and 'idiot' were originally introduced as technical terms to do much the same job as describing those who are mildly, moderately or severely-profoundly retarded. The expressions 'retardate', 'feeble-minded' and 'mentally defective' are no longer acceptable today.[2] Similar points could be made about other languages for this is not just a feature of English. Rather, it shows language is a subtle indicator of social mores.

L. Kopelman and J. C. Moskop (eds.), Ethics and Mental Retardation, 65–85.

Summary. I will defend the view that all humans merit some respect as fellow beings. Defense of this, however, requires clarification of what is here meant (and equally important what is not meant) by 'respect'. For to apply the term so globally indicates it cannot mean esteem or entail that the individual respected must function in a certain way. In contrast, there is a tradition which bases respect and rights on individual capacities thereby excluding some very retarded humans as beings who merit respect and rights. As this seems to me to be mistaken, and both views cannot be true, particular attention will be paid to the most severely retarded in justifying the position that all humans as fellow beings merit a respect of sorts. I begin by illustrating that we do not always think it reasonable to base rights and respect upon the kind of capacities or potential human individuals possess. Consideration of how we *justify* the judgment that there is a handicap illustrates that there is a kind of respect which is shown to fellow beings that is quite independent of their degree of function. To justify the values and goals used in estimating whether or to what degree someone is handicapped is to show them respect. To make a judgment in a cavalier fashion shows disrespect whatever the capacity of the human being evaluated. It is argued in *Section I* that respect involves justifying the goals of labeling by showing it (i) benefits the individual, (ii) benefits society, or (iii) is needed to maintain the integrity of observation and description. The goals used must be clarified and defended because there is a potential of harm in ascribing handicaps and because these reasons can come into conflict. In *Section II* it is argued that this important kind of respect is required to justify labeling and the use of values in describing those who are handicapped (whether by estimates of I. Q. level, adaptive behavior or developmental level). These judgments are necessarily value laden in two ways. First, standards are used that are evaluative; and second, justifying the use of the label creates certain obligations. These standards and obligations are unaffected by the client's level of functioning. This is an important example of a central point I wish to make; namely, that respect for and obligations to someone are not always grounded in that human's capacity, agency or potential for reciprocity. If we ought to show certain respect no matter what the degree of retardation, then we need to say what sort of respect is meant. In *Section III* four kinds of judgment about respect are distinguished: 'esteem', agency', 'status' and 'limitations'. This serves as a basis for addressing the central issue, raised in *Section IV: How can the established (and I believe correct) conviction be accounted for that humans, even those profoundly retarded, are owed respectful treatment or respect*? To justify this conviction respect can be based on important features of their *status as fellow-beings*. That most people do show them respect does not prove they ought to. To show it is rational to respect them we will focus on the importance of what they *share* with us: (1) They, like us, are beings who are sentient; thus, we respect them as fellow beings who feel; (2) They, like us, are individuals whose treatment by us affects our institutions; thus, we respect them out of our own interests; and (3) They are members of our communities and families; thus we respect ties of community, affection or benevolent concern in a way that goes beyond the minimal requirements of (1) or (2). It is argued that moral theories that base respect, rights and obligations on individual features alone (such as on sentience or on potential or actual moral personality) are in a difficult position when their theories are applied to considered judgments about how to treat humans who are very retarded.

I. JUSTIFICATION OF ASCRIPTIONS AS A WAY OF SHOWING RESPECT

Often 'labeling' is used pejoratively as if it is part of the handicap people with handicaps must suffer. We sometimes need, however, to describe individuals including their intellectual capacities. Frequently labelling others as 'handicapped' benefits them by signaling that they have important and unique needs. It can be *very* useful in making special care, programs or funds available to them; in defining roles and responsibilities; in determining what reactions are or are not appropriate; in showing the need for new programs, facilities, screening programs, research, or laws. Thus, the first reason for labeling others is *to benefit the person* and it is a most important one in a professional setting (medicine, psychology, nursing, social work, etc.). Still, labeling needs to be done with care, for singling out or labeling persons can sometimes cause harm through stigmatization, loss of self-esteem, assignment of a disvalued role, self-fulfilling prophecies or prejudicial treatment. Thus, if benefit to the person is given as the justification for labeling by those with special authority in our society, such as physicians, psychologists or teachers, then the person has a right to be guaranteed that on balance the name or label will be likely to produce more good than harm for him [22]. Consideration of harms should include: psychosocial dangers already mentioned; misconceptions (e.g., retarded people are more dangerous than others); irrational attitudes (e.g., thinking of disability as punishment from God); and "spread" effects, or the tendency of disabled persons and others to view them as more handicapped than they are [17] [31].

A *second* reason for labeling others as handicapped is for *benefit to society*. People who have uncontrolled seizures or cannot understand road signs are not permitted to drive for the good of the rest of us. To justify labeling or restricting others for the sake of social utility should require a heavy burden of proof. The exact nature of the harmful effects to be avoided or the beneficial consequences to be achieved and the probability of their occurrence as a reasonable basis for the restriction must be demonstrated [16].

A *third* reason for labeling is for accuracy or for *integrity of observation and description.* For example, a physician needs to include on the medical chart that an infant has phenylketonuria (PKU) or the Trisomy 18 syndrome. Such descriptions may lead to help for the patient, for his family or, through research, for future generations of patients. And, where the disability is apparent or extreme, a diagnosis *per se* probably will not harm them. Sometimes, of course, they can harm. For example, some untreatable conditions like XYY syndrome, where there is great potential for psychosocial harm, should not, in my view, be in nonmedical or readily accessible records even though it is an important observation about the individual; and some research which might benefit society should not be done out of consideration and respect for the rights and welfare of these persons [16]. In general terms, this means that these three interrelated reasons for labeling – (i) benefit to the person, (ii) social utility (including research), and (iii) integrity of observation and descriptions – can occasionally come into conflict.[3] Furthermore, they use values and create obligations that must be justified.

II. THE USE OF VALUES IN CALLING PEOPLE HANDICAPPED

In this section I will argue that labeling others as handicapped, no matter what the kind or degree, uses and creates values or obligations. If this is correct then there is no value-neutral way to describe people as handicapped and no escape from the moral responsibility of justifying these values or purposes of the ascription. This illustrates that rights, values and obligations do not always depend upon the degree of incapacity. For example, one has the same obligation to be as accurate as possible if the retarded person can use language or not. It will be argued that judgments that people are retarded or handicapped are necessarily value laden. This is as true of I. Q. tests as it is of other tests of functional, adaptive or developmental levels.

In contrast to my view, however, it is sometimes argued that retardation can be defined or ascribed in a value-neutral manner. Some have argued that the statistical-frequency model results in value-free testing [20]. According to this method, distribution of a trait, in this case intelligence, is determined on the basis of certain standardized tests. Those within one standard deviation of the mean (68%) are judged to be normal; those one to two standard deviations from the mean (27%) are judged high or low normal; and those two or more standard deviations from the mean (5%) are judged abnormally low or abnormally high. This model has been used by the American Association of Mental Deficiency to define mental retardation [20]. Using the Wechsler Scales, Stanford-Binet and Cattell tests, this association defines the mildly (2–3 SD), moderately (3–4 SD), severely and profoundly (4–5 SD extrapolated or estimated) retarded.

Jane Mercer argues that the statistical-frequency model is value neutral because it may be good or bad to be in the high, low, or average group [20]. This does not show that tests or results using this method are value neutral, but that different norms and values may be used to define the groups or interpret the result. A first difficulty with the view that values are avoided, then, is that it must initially assume that the statistical norm or average indicates what *ought* to be the norm for judging. And this is not only value laden, but requires justification. We would not think a good intelligence test to consist, for example, in how loudly one can scream, in one's physical strength or in one's tendency to exhibit perseveration. Thus, values are introduced in the selection of norms judged appropriate for the purpose of the tests.

Moreover, additional non-statistical or normative considerations come into play in the selection of one test instrument as more apporpriate than another. This is obvious when there are disagreements about whether or not tests ignore important parameters for judging, or artificially compress key features into a single measure so a scale can be established [20, 26]. Such debates about the sensitivity of the tests used to estimate the range of talents people have presuppose values, namely, that we can judge that some test instruments are *better* than others either in general or for certain purposes. For example, many recommend that scales based on measurements of adaptive behavior or functional level should supplement or replace standard psychological testing to fix I. Q. levels for individuals. Methods are not developed because they are considered worthwhile or desirable in themselves, but because they serve as tools or a means to certain ends, goal or values.

The selection of norms and tests employed, the uses to which the tests are put, and the interpretation of the results, all introduce nonstatistical and evaluative considerations. For example, consideration should be given to the tests' fairness in judging those from different cultural backgrounds, or where poor performance may be due to fatigue, hunger, or social or economic deprivation. Mercer directly raises these issues of respect and fairness, charging that in the United States patterns of Anglocentrism in labeling the mentally retarded can be found in schools and other organizations. Thus, she inadverently shows that, far from being value free, these tests require careful moral justification including their fairness; for even if a test is a good measure when applied to some groups, it may be unfair for others.

Furthermore, the results obtained can mean different things for different groups. In order to interpret the results we need to distinguish between the statistically most frequent norm, and the norm which represents an ideal or what we judge ought to be the norm or ideal for the population being tested.[4] The extent to which the average and ideal converge cannot be determined by the statistical method itself [14]. The same bahavior, e.g., crying, can mean different things depending on age or culture and can, therefore, elicit very different interpretations. Judgments about what ought to be the norm include how well people do or are likely to function; and these estimates certainly are not value free. For one thing, the boundaries between normal and abnormal function can be drawn differently based, in part, on what is valued or approved. While great impairment of function, such as that resulting from Tay-Sachs Disease, would warrant ascribing disability in any culture, some ascriptions are not generalizable cross-culturally. There is a wide range of adaptive and nonadaptive functions involving many parameters such that not all will view the same things as nonadaptive. A person unable to function independently in a highly technical society might do very well in a stable rural community.

Standardized measurements use criteria that are out before us for critical review and testing. And this is obviously an advantage over informal estimates for describing intellectual abilities. Though none are value free, different methods have different advantages and disadvantages. A common and important way to describe the ability or potential of retarded persons compares their capacity at maturity to that of the development of normal children. Typically these standards are also well-defined and established by the statistical frequency model. For example, one might hear that "Adults with Trisomy 21 typically reach a mental age equal to that of a child of eight". Like the I. Q. test, this is useful and helps us know what to expect. Moreover, such expressions compare retarded persons to children, a group to whom we feel very loving. However, there are many things retarded adults with a developmental age of eight can decide for themselves that eight-year-old children cannot. This way of expressing ability, while important, has dangers; it reinforces an association of retarded persons with children or as childish, thereby increasing the risk of ignoring their need for respect as persons who set their own goals and make their own plans.

Thus, the language, definitions and measures we use are shaped by our attitudes, values and interests. Describing the intellectual abilities of retarded persons is evaluative, because judging people to have intellectual or other handicaps means we believe that they *fail to meet certain norms or ideals considered good, desirable or worthwhile*. Not only are they value laden in this *narrow* sense, but also in the *broader* sense that *these ascriptions create obligations* [10]. For example, physicians or psychologists who make these judgments have obligations to help their clients if they can and to see that the labels are accurate and current.[5] This includes using well-tested and multiple measures and checking for other factors which might give a mistaken impression about intelligence, such as depression, or economic or social deprivation.

Decisions that people are handicapped, then, are not value free. The temptation to say they are seems to come from good, though misguided, intentions. It seems supposed that if we do not admit we think of them as failing to meet some standard, we do not judge them. But respect is shown in the moral justification of why and how the estimate is made and not by concealing its nature. The sentimentality or pretense that denies we judge them as handicapped seems dangerous to me because it may also be used to deny the need for the moral justification of such judgments.

Generally speaking, those who make such estimates seem conscientious and more sensitive than most of us about the impact of labeling on people's lives. My purpose has not been criticize them or the tests but to show that the value-laden character of judgments about retarded people raises important issues about rights, value and responsibilities which deserve critical attention.[6] It has been used to illustrate that some respectful treatment we show our fellow beings does not depend on their individual capacities. If we say that all humans deserve respect no matter the level of their intelligence then we need to say what we mean by 'respect'. In the next section several kinds of respect will be distinguished to determine which is most helpful in understanding this. The kind of respect some ethical theories have selected as basic excludes the severly retarded as beings who merit respect. I object to selecting it as basic and suggest it may affect our judgment of the plausibility of these theories.[7]

III. FOUR KINDS OF RESPECT

This discussion of language and labeling has focused on the attention, regard or consideration that we ought to give retarded persons. As such, it has been implicitly related to one of the most basic notions in ethics, that of *respect.* I now wish to make that relationship more explicit. We have an opportunity to show respect (or disrespect) by the way we employ language, acknowledge obligations, are concerned about the accuracy of diagnosis or labels, and give regard to the impact of our actions on their lives. Most, *but not all* retarded humans are agents and use language, make plans, have goals, show compassion, have a sense of what is and is not fair, and are pleased by their achievements; they, like the rest of us, seek respect, enjoy the esteem of others, prefer the chance to make decisions, and wish for acceptance.

If we ought to show some respect to all humans, no matter what their talents or potential, then we need to say what kind of respect is meant. In order to clarify the notion of respect with special reference to the *varied group* of persons we regard as retarded, several different, though related, ways of using this term will be compared:

(1) I respect the work of Richard Diebenkorn.
(2) We ought to respect her ability to make her own plans.
(3) I respect him in his capacity as a United States Senator.
(4) We respect this disease enough to take precautions for prevention and to seek a cure.

For convenience I will call these different senses (1) *esteem*; (2) regard for *agency*; (3) regard for *class membership or status*; and (4) attention to or acceptance of *limitations.* After distinguishing these different senses, I will argue that it is the *third* sense of respect, regard for status, that will be especially helpful in understanding and justifying our belief that all humans as fellow beings merit respect, even profoundly retarded human individuals.[8]

(1) *Esteem.* By using respect in the sense of honor or esteem, we offer a judgment about the virtue or worth of ends or means, or of something or someone's character traits, actions, or what intentionally flows from them. It can be used in moral as well as aesthetic judgments such as, "I respect the courage she showed." Occasionally we indicate a more general or unqualified esteem for the kind of life someone has led, as in "I respect Dietrich Bonhoeffer;" but usually our approval is limited in some specific way. This qualified or elliptical sense singles out some feature for approval; it is very clearly used, for example, when we express admiration for an enemy's skill. We sometimes change our minds about these normative judgments; for we distinguish between the sincerity of and justification for esteem, and rec-

ognize our fallibility in making these appraisals. And when we care to (say there is a prize at stake), we can discuss and evaluate our reasons for singling out someone for special honor. People usually earn or inspire this sort of respect, as Brother William apparently did.

We distinguish character traits, voluntary acts or dispositions to act, from basic abilities. We want to say that one is not responsible for most physical attributes and intellectual capacities in the same way one is responsible for voluntary acts of bravery, kindness or diligence. The former are the cards nature has dealt us but the latter reflect how we play them. However, the distinction between them is not sharp because how we play our hand depends on what we have been dealt.[9]

We honor people for being clever inventors, talented playwrights, brilliant lawyers, knowing that their accomplishments take special natural abilities or talents that were used well and for good ends; but we also honor people for being kind, generous, diligent, courageous, compassionate, etc., knowing that beyond a certain level, these acts are not tied in any obvious way to being innately gifted with intellectual or other natural capacities. Insofar as esteem presupposes voluntariness it could not apply to beings incapable of voluntary acts. Thus, as profoundly retarded individuals are incapable of voluntary acts, it would not make sense to esteem or blame them for what "they do." Such judgments would be inappropriate as they presuppose conditions that cannot be fulfilled. At minimum what is required is a sense of self or of time, to have hopes, beliefs, or goals, or to foresee the likely consequences of a decision. The second sense of respect presupposes the capacity to make choices, thus it is inapplicable to beings incapable of voluntary acts.

(2) *Regard for Agency.* Another sense of respect honors the ability of others to make decisions for themselves and control their own destinies. It presupposes that those who act do so deliberately; that they believe they know what they are doing or mean to accomplish by their act. In contrast to things that just happen to us, agency marks a capacity for purposively initiated action. *Respect for persons* in this sense of regard for their dignity as agents or potential agents has been emphasized in many major theories from those of Locke and Kant in earlier times to Nozick and Rawls today. Murphy points out that respect for persons in these theories functions as a technical philosophical expression; it is used to stress the special regard that should be shown to beings, persons, who are sufficiently rational to control their own destinies [22]. To treat others with respect in this sense is to be willing to

grant they have independent views, aims and interests. Moreover, it is a willingness to offer, in good faith, reasons for what we do when others are affected by our choices, or when we must overrule the choices of others ([27], p. 337). We all want to be taken seriously and treated courteously. Few if any of us could withstand a uniform assault on our self-esteem from others who regard us with indifference or contempt or our goals as trivial or deserving of ridicule. Showing respect for each other in this way fosters our mutual need for cooperation and support. Everyone benefits from living in a society where the duty of mutual respect is honored ([27], p. 338). This sense of respect, regard for the agency of persons, serves in many theories as a basis for negative rights or liberties in contrast with positive rights or entitlements [22, 27]. Interference with the choices or destiny of a competent person for benevolent or paternalistic reasons is generally viewed as disrespectful, an offense to their dignity and status as rational beings [15, 22]. *Thus the second sense of respect, agency, could not apply to beings incapable of voluntary actions. If the most profoundly retarded humans are incapable of voluntary acts it would not make sense to show regard for them as agents.*

Most retarded persons are capable of voluntary action and, like the rest of us, seek the approval of others. But the approval or respect of others for what we do (in the first sense of esteem) presupposes respect or recognition by others of our ability to act independently and deliberately on our own behalf (the second sense). Moreover, it presupposes the chance to exercise that ability in order to have the opportunity to earn approval so that native ability or potential will develop and not atrophy.

One of the practical difficulties in dealing with retarded adults who are not fully competent yet are capable of some voluntary actions, is estimating the extent to which they are capable of making decisions about themselves, for themselves. For some, paternalistic obligations (others acting to benefit them) ought to dominate and it should then be up to competent adults to decide how best to protect them and to foster their development. However, many retarded persons deserve respect as persons capable of autonomy (expressing their true natures in decisions for themselves). There are practical problems, of course, in determining the degree of responsibility and protection that is best for each person who is not fully capable of caring for himself. Sometimes paternalism and autonomy conflict as in the case of older minors at odds with parents. And consideration of older minors is an instructive model, as most retarded persons are mildly retarded. But frequently paternalism fosters autonomy. That is, having competent adults making decisions for those who are not competent adults can encourage their development. A seven-year-old cannot be allowed to express himself by not taking baths or not going to school. But it is a complex relationship; for sometimes, in allowing persons somewhat more responsibility than we think they have earned, we encourage autonomy or recognize that our estimates were mistaken [2].

Whenever it makes sense autonomy ought to be encouraged. Of course, one might object that "we ought to encourage autonomy" is a fine but ineffective policy because it allows for such wide interpretation. While this is not trivial as a practical objection, its over-emphasis may cause one to miss the point. For, this is a recommendation for a certain attitude about how people ought to be treated. Its value is to favor a conceptual framework that seeks to encourage (e.g., by laws, policies), rather than discourage, independence and autonomy "as much as it makes sense" or "as much as it is possible." This sort of attitude can be as potent in determining how retarded people will be treated as another attitude which simply assumes all retarded persons are always best off having guardians make all decisions for them.

(3) *Regard for Class Membership or Status.* There is another kind of respect which differs from esteem, or from the regard for agency. It is the kind of respect intended when we show regard because an individual occupies a certain rank, role, job, or is in a certain relationship or circumstance. The Biblical commandment that children shall honor their parents indicates that a certain respect or dutiful regard is due to parents because of a place they occupy in our lives. We may hold them in esteem as well but that is not what is commanded, because usually that must be earned or inspired. Showing respect in this sense leaves open whether or not I like or approve of the person owed the respect. I may dislike someone, yet still feel inclined or obligated to show regard or courtesy to him as a judge, teacher or senator. What I respect in these cases is the person's role, position, office or status within certain social institutions. Sometimes one simply has respect for the power of the person's office; and this is close to the fourth sense of respect.

(4) *Attention to or Acceptance of Limitations.* We can respect thunderstorms, bombs, diseases, locusts and good things like needed sunshine or rain; and what this means is that we give special attention or regard to their power in our lives. Correlatively put, it acknowledges and accepts our limitations to do certain things. When battles with enemies, elements, diseases and pests are won, we feel encouraged by our skill, endurance or inventiveness in overcoming limitations. However, like thunderstorms, it is sometimes beyond our

control to prevent the conditions causing retardation. When we say someone is *retarded* we mean we consider the underlying condition to be essentially *irreversible* [2]. We do not say the infant with PKU is retarded until we think the irreversible, permanent damage is done. Of course this does not rule out working for improvements or seeing that our judgments are fallible; but this is what we mean by using these words. To a society and health system focused more on *cure* than *care*, retarded persons can make us uncomfortable. For they remind us that there is a tragic side to life, that there are events evoking pity and fear because they are beyond our control.

It is important to see which of these kinds of respect are applicable to the claim respect is owed to all humans. The first sense of respect, esteem, is won by relatively few retarded persons outside their families. However, regard for agency, or the second sense of respect, applies to most retarded persons because most are capable of voluntary actions. The sense of limitations, the fourth sense of respect, applies to us all. It indicates our limitations and why we think we ought to try to accept and care for what we cannot prevent or cure. *But it is the third sense, regard for status as beings of a certain kind, that is especially important in giving an adequate account of the established, and I believe correct, conviction that even the most profoundly retarded individuals are owed respect.*[10]

IV. RESPECT FOR ALL FELLOW-BEINGS

To summarize, our goal is to justify the claim that all humans are owed a kind of respect. But respect here needs to be understood as what we share that enables us to function as fellow beings. We will now attempt to give a positive account of this. It has already been shown, by means of the labeling analysis, that some important kinds of respect do not depend on individual functioning. We have discussed the obligations arising from assessments that others are a retarded and the rights of those so judged. It was argued that these rights and obligations do not always depend on the level of intelligence or function of any individual judged. Even if those who are profoundly retarded are unable to exercise their rights, their rights are not dissolved nor are obligations to them wiped away. This was illustrated by our discussion of labeling. In this section I will pursue this topic on a more theoretical level by arguing first that ethical theories attempting to tie respect or rights to some minimal level of intelligence or functioning are problematic. And second, focusing on that which humans just because they are fellow beings share illuminates why all are owed certain things and have certain rights. This view that all humans are owed respect does not say just what form this takes. For example, it means that human fetuses should be more importan' than rat fetuses to us but entails no anti-abortionist view; it does not say how a society must support each profoundly retarded individual, but it does mean they should not be made to suffer unnecessarily. No specific programs are here proposed in this discussion of why these claims are reasonable.

I will argue that there are three reasons which taken together justify the conviction as reasonable that all humans, even those profoundly retarded are rights-bearers and individuals who are owed respect as fellow-beings. What they share with us is important enough to merit respect. These three reasons, though interrelated, focus on different things about what we *share.* Each of these reasons is related to sense *three* above, *regard for class membership or status.* First, they share a capacity to feel and their *sentience* should be respected. Second, as our discussion of labeling illustrated, how they are treated affects our institutions; thus, it is in our own *self-interests* to see that they are treated respectfully. Third, beyond the minimal requirements of sentience and self-interest, we share our communities and homes with them; we respect the *commitment, benevolent concern or affection that holds families and communities together.*[11]

The sharp contrast to my view are those positions which assert *individual features alone* can serve as a *basis* for understanding and justifying rights and obligations to human beings. One of these is that the only reason for showing regard or consideration to the profoundly retarded is that they feel *pain and pleasure.* But, while important, to say that it is the only reason fails to account for our reasoned conviction that we have obligations to profoundly retarded individuals (or for that matter permanently comatose ones) different from those we have to rats or worms. Animals feel pain and pleasure and it is wrong to make any sentient creature suffer needless pain. To argue from this to suppose they deserve the *same* sort of consideration as our fellow humans ignores the social relations humans in communities acknowledge, perhaps necessarily, to each other that they do not have to most animals [5, 9].

Another view, which also seems to me incorrect, is that the only reason we extend consideration, care or rights to profoundly retarded individuals is *for ourselves or reasons of self-interest.* This view arises in an interesting way in Rawls' theory of justice. Stressing the second kind of respect (regard for agency), Rawls adopts the view that respect should be given to beings who have moral personalities. By this he means all those with a conception of their own good and who are capable of a sense of justice ([27], p. 19). He regards moral personality rather than the capacity for pleasure and pain as the fundamental aspect of the self ([27], p. 563). This would explain why showing respect for persons as moral beings involves trying to understand their aims, interests, and point of view. Furthermore, it shows why our willingness to give reasons for our actions that affect others is an appropriate sign of our respect for them.

The overwhelming majority of human beings, and this includes most retarded persons, have the capacity for moral personality and this is sufficient to secure for them the full guarantees of justice and consideration. Even infants are guaranteed justice and consideration in society by virtue of their potential, where, in the normal course of events, they will mature and become independent and autonomous persons. I agree that moral personality (or the capacity to develop it) is *sufficient* for inclusion in the category of beings who have rights and to whom we have duties. However, Rawls needs to consider the consequences of his view for the small percentage of human beings who show no evidence of, or potential for, developing into beings who have a moral personality (i.e., beings who have a conception of their own good or who are capable of a sense of justice). Some cannot communicate and seem not to have a sense of self or time.

If, as Rawls says, we owe equal justice to individuals for reasons of their moral personality, and if we agree some profoundly retarded persons do not have the capacities or potential required, then there is no obligation to treat them with strict justice. Rawls seems uncomfortable with this implication, for he admits that "those more or less permanently deprived of moral personality may present a difficulty" for his view ([27], p. 510). He relies upon a *self-interest* and *pretense* – pretending the condition has been met – to try to justify why we regard profoundly retarded individuals as entitled to consideration, justice and respect even though they show no capacity or potential for moral personality:

> Whether moral personality is also a necessary condition I shall leave aside. I assume that the capacity for a sense of justice is possessed by the overwhelming majority of mankind, and therefore this question does not raise a serious practical problem. That moral personality suffices to make one a subject of claims is the essential thing. We cannot go far wrong in supposing that the sufficient condition is always satisfied. Even if the capacity were necessary, it would be unwise in practice to withhold justice on this ground. The risk to just institutions would be too great ([27], p. 506).

Rawls' answer, then, is that we pretend they have moral personality and we do this *for ourselves* and *not for them* [22]. We do not want to run the risk of undermining just institutions for the sake of a very few individuals. For making or enforcing judgments about who does or does not have moral personality would be too risky.[12]

To those who argue that there is a *lack of reciprocity* (they cannot give back what they get) when respect is extended to the profoundly retarded, Rawls answers with powerful considerations based upon self-interest; indeed his view is elegant in singling out concerns based on self-interest as a basis for justice. But as Joel Feinberg and Jeffrie Murphy have pointed out in criticizing Rawls, it is highly counterintuitive to suppose that the only reason we have in extending protection and the duties of respect to others is our own self-interest. They argue that protecting the sensibilities of others has a prima facie claim to satisfaction, so that we also do so for their sakes [22, 7]. In addition, however, we need to consider relations between human beings. For, even if self-interest goes a long way to answer the "what is in it for me?" question about showing regard for those who do not have moral personality, it fails to give sufficient attention to relational characteristics between humans; it fails as a full account for why we think we ought to support or help our fellow human beings.

In contrast, I want to argue that there is no one feature of individual human beings (such as moral personality or the capacity to feel pleasure or pain) which entirely explains this, or marks us and only us out as special beings deserving a certain sort of respectful treatment.[13] Why then should we respect each other in a way different from nonhuman animals (worms)? We do recognize that even those who do not show moral personality, intelligence or use language are owed certain things. We think for example that where ever possible, they ought to be warm, dry and housed.

We depend on each other for support and protection in a way that makes or should make the interests of all human beings of special importance to us; more important, say, than protecting the interests of nonhuman animals. Biological and social relationship creates unique attachments and felt obligations. Using common language and sharing economic and social arrangements bind us together. We accord special significance to the interests of other human beings, not because each and every one of us has some special capacities or potential but because we stand in certain relationships to each other as human beings [5] [9]. The notion that we inspect human beings for certain necessary traits, such as sufficient potential for rationality or intelligence to determine if they are worthy of consideration and care seems implausible when we consider social and family relationships, friendships and affection.

Though they disagree on what form it should take, members of our communities with retarded family members often are among those who remind us of the importance of showing consideration and respect to those who are intellectually handicapped. In our concern for them and their relatives, we want to support them. Of course, the regard people show can take different forms, like lobbying for better screening, diagnosis, funding, facilities, training or support for studying causes or possible cures. In the case of those permanently devoid of capacity or potential, those profoundly retarded, there is a lot of disagreement in our society over what constitutes respectful treatment. Some believe that, under certain circumstances, respect, consideration and care are best expressed by withdrawing or withholding therapy where the prognosis is very poor and the therapy would be disorienting, painful, and frightening. Sometimes alleged expression of regard or respect can take forms which are far more controversial than this. Some elderly parents of retarded people have thought killing them showed greater respect then sending them to the institutions available to them. This sad example says something about these institutions, or at least that the sincerity of the respect is different from justifying something as a respectful practice.[14]

To understand the basis of respect in the sense of acknowledging obligations to *all* human beings, including profoundly retarded ones who cannot act or reciprocate, we need to move beyond characteristics just describing individuals, self-interest and include descriptions of social relations, benevolence and affection.

The importance of regarding each other as fellow beings can be illustrated by means of a fanciful example that would have us regard ourselves as relatively low rather than relatively high on the intelligence scale. Suppose a modern-day Ponce de Leon found a fountain of intellect. By drinking from it, individuals could greatly increase their intellectual capacity and become many times more intelligent than Plato, Shakespeare or Leonardo da Vinci. Greater intelligence is instrumental in advancing all sorts of purposes, bad as well as good. So, if excluded from drinking at the fountain of intellect, some might want to destroy it; some might want to control who had access to it, stipulating that only very virtuous "incorruptible" people could drink from it. (That should rule out anyone drinking from it.) Our concern would not only be due to their increased power over us, but also due to the likelihood that those who drank from it would come to regard us as creatures unworthy of their respect. That is, by virtue of our different status and limited capacities, they would regard us as inferior beings. They might agree to show some regard or to respect us as creatures capable of suffering. As we are creatures who feel pleasure and pain, they might agree that *some* duties of compassion apply. They might praise the simple tasks we do, the easy problems we solve, and recognize our limited capacity for vice or virtue. They might make pets of us or assign us "guardians" who had authority over everything we do.[15] They might say, however, they cannot respect us in the same way as rational beings as they do the rest of their kind because of our very limited natural capacities. Suppose we could not rebell or escape. Then, we might be at their mercy, dependent on their kindness and compassion, and hopeful they would allow us the opportunity for some autonomy and for the respect we believe we deserve. The lack of reciprocity would be apparent and, if the effects were permanent and could be passed on to their offspring, they might disown us as members of their families and as fellow beings saying we do not share enough with them. However, if the effects were only temporary, or if the greater intellect could not be inherited by their offspring, they might still identify with our suffering, want our cooperation and good will, and think of and show regard for us as fellow beings. That could be the key, and our best chance for respect. From our disadvantaged position, of course, *we* would not be convinced that unequal abilities, or lack of reciprocity, justifies abandoning respectfulness to us.[16]

Conclusion. I have tried to show that all human beings merit respect based on their status as fellow-beings. One might charge that all I have shown is that most people do recognize that we ought to act respectfully to the retarded, but that what we need to show is that this is rational. But I have tried to show this is reasonable by considering in detail the moral requirements of justifying the ascription of handicaps. We do not consider the level of functioning in fulfilling these moral requirements. Not only do we show respect by requiring a moral justification of the norms we use, but in applying them fairly to all regardless of their ability.

To show respect to someone does not always require esteem, reciprocity, affection or their ability to do certain things. What remains is a kind of respect or courtesy that is neither trivial nor sentimental. It is shown in facing up to the moral responsibilities engendered by justifying the reasons why and how we judge that others are handicapped. And it is demonstrated by serious discussion of what it means to consider the interests of all humans or to show respect to special groups. Critical reflection upon how the retarded ought to be treated will challenge many views which have never really been considered in relation to this sizable minority of our fellow human beings. I have argued that some ethical theories will have great difficulty in accounting for established, and I believe correct, convictions that even human individuals not showing evidence of personality, voluntary action, or a sense of time or self, nonetheless have certain rights and are owed certain things.

East Carolina University School of Medicine
Greenville, North Carolina

NOTE

[1] Brother William was briefly discussed in a lecture by the Late Professor Milton Williams of Syracuse University.

[2] Good intentions aside, these terms seem to have a "half-life" of acceptability. I wonder how long the now current terms 'exceptional' and 'special' will last, or how long Professor Spicker's 'atypical' would survive were it to become popular. It is not the words but the underlying attitudes that need changing.

[3] Professor Spicker assumes it is my view that these three interrelated reasons for labeling are justified if they "engender more social good than ill" ([29] p. 87). But this position does not seem obvious, would have to be argued for and is certainly not defended here. For example, on some crude "cost-benefit" type of argument one could say labeling certain individuals might benefit those labeled, but open up so many costly entitlements to them it might not be in the public interest.

[4] There is also a problem created by shifting norms in populations.

[5] In some cases, such as PKU, Trisomy 18 syndrome or Tay-Sachs disease, it is very clear what standards are being adopted and why. However, other labels are infamous for their lack of clarity. For example, minimal brain damage (MBD) is recognized to be a catchall diagnosis – vague, overused and often misapplied. Interestingly enough, there are a variety of names, more or less fasionable, for this ill-defined condition(s). Changing the names, of course, will not in itself clarify the criteria for its use; however, it might illustrate an earlier point about shifts in language.

[6] This has been discussed peripherally by philosophers in the animal rights literature.

[7] It will lead us to consider a range of problems about how we should treat others which, though important, have largely been ignored in the philosophical literature.

[8] There are other ways 'respect' is used. My list is not meant to be exhaustive, nor does it indicate the full extent to which these senses overlap or are normative .

[9] Hume argues that virtues and abilities cannot really be distinguished.

[10] In reply to Professor Spicker, it is the third, not second kind of respect that I stress is important to understand why consideration and care should be extended to the humans we call 'retarded'. Agency (the second kind) and esteem (the first) presuppose beings capable of voluntary acts. We argee that some profoundly retarded *individuals* are not agents, not persons capable of actions and may not show an awareness of time self or personality. Yet there is a conviction they merit respect of a sort. Professor Spicker seems to agree. My goal is to attempt to account for this in terms of the third, kind of respect, regard for their status (1) *as* sentient beings (we do it for them), (2) *as* beings who affect our just institutions (we do it for self-interested reasons), and (3) *as* they are our fellow human beings (we do it because we are social creatures). Focusing on the second kind of respect naturally leads to the problem of reciprocity. My solution, I believe, avoids this problem. Professor Spicker, by distinguishing between *attitudes of respect* and *principles of respect*, seems to adopt the view I wish to attack. For it suggests (A) that respect should be interpreted along the lines of Kantian and Rawlsian theories such that moral personality and reciprocity are basic to all rights; and (B) that inward feelings and outward actions to others are more sharply distinguishable than they are in fact. However, Rawls admits he has a problem because profoundly retarded non-agents may have rights that cannot be accounted for in these ways. And Spicker gives no reason for claiming that moral personality and reciprocity are always basic in this way. In light of the rights given and duties people adopt to very retarded individuals, this view of Spicker's seems unsupported.

[11] We must attend to certain social or relational characteristics between our fellow human beings [5, 9]. This final consideration is largely ignored by ethical theories which base rights and obligations on individuals' possession of certain characteristics, such as rationality, moral personality, intelligence, using language or feeling pain or pleasure. It may be the third could be stated to include the other two; or that they should simply be stated in terms of egoism and altruism. But my purpose is simply to make certain features salient which are used to explain this conviction.
[12] A variation of Rawls' view is that we show respect to profoundly retarded individuals on the grounds it is *logically* possible for them to develop moral personality. But this is inadequate. For, if granted on this basis, then there is no reason to exclude anything which could be logically conceived to develop moral personality – including non-human animals, plants, etc. Even though Rawls' position is troublesome here, it is better off then Nozick's. Nozick ties moral personality to capacity to control one's life, so similar criticisms could be offered of his position. But Rawls, unlike Nozick, recognized mutual aid and mutual respect as natural duties.
[13] Transcendental characteristics that have been attributed to persons, like being children of God or members of the kingdom of ends, are not discussed.
[14] I do not discuss the issue of resource allocation or the limited resources of society. As I mentioned, regard and consideration can take many forms and I do not say they are the same for all societies.
[15] Parallels to the animal rights literature are clear. I found most people would want to be brighter but not so much that they would feel isolated. If this is so, it makes a similar point.
[16] This shows the importance of taking moral personality as a sufficient condition.

BIBLIOGRAPHY

[1] Aristotle: 1941, *Poetics,* in Richard McKeon (ed.), *The Basic Works of Aristotle,* Oxford University Press and Random House, New York.

[2] Bedau, H. A.: 1967, 'Egalitarianism and the Idea of Equality', in J. R. Pennoch and J. W. Chapman (eds.), *Equality,* Atherton Press, New York, pp. 3–27.

[3] Benn, S. I.: 1967, 'Egalitarianism and the Equal Consideration of Interests', in J. R. Pennoch and J. W. Chapman (eds.), *Equality,* Atherton Press, New York, pp. 61–78.

[4] Brandt, R.: 1970, 'Traits of Character: A Conceptual Analysis', *American Philosophical Quarterly* 7, 23–37.

[5] Diamond, C.: 1978, 'Eating Meat and Eating People', *Philosophy* 53, 465–479.

[6] Dworkin, G.: 1976, 'Paternalism', in S. Gorovitz, *et al.* (eds.), *Moral Problems in Medicine*, Prentice-Hall, Englewood Cliffs, pp. 185–200.

[7] Feinberg, J.: 1974, 'The Rights of Animals and Unborn Generations', in W. T. Blackstone (ed.), *Philosophy and Environmental Crisis*, Athens, University of Georgia Press, Georgia, pp. 43–68.

[8] Foot, P.: 1967, 'Moral Beliefs', in P. Foot (ed.), *Theories of Ethics,* Oxford University Press, London, pp. 83–100.

[9] Francis, L. P. and Norman R.: 1978, 'Some Animals Are More Equal Than Others', *Philosophy* 53, 507–527.

[10] Frankena, W.: 1967, 'Value and Valuation', in P. Edwards (ed.), *The Encyclopedia of Philosophy,* Vol. 8, Macmillan Publishing Co., Inc. and The Free Press, New York and London, pp. 229–232.

[11] Hare, R. M.: 1972–1973, 'Principles', *Proceedings of the Aristotelian Society,* 73, 1–18.

[12] Hauerwas, S.: 1977, 'Having and Learning How to Care for Retarded Children: Some Reflections', in S. J. Reiser, A. J. Dyck, and W. J. Curran (eds.), *Ethics in Medicine*, M.I.T. Press, Cambridge and London, pp. 631–635.

[13] Hume, D.: 1953, *An Enquiry Concerning the Principles of Morals*, Open Court Publishing Company, LaSalle, Illinois (originally published 1777).

[14] Jahoda, M.: 1958, *Current Concepts of Positive Mental Health,* Basic Books, New York, London.

[15] Kant, I.: 1959, *Foundations of the Metaphysics of Morals*, (transl.) L. W. Beck, Bobbs-Merrill Co., New York (originally published 1785).

[16] Kopelman, L.: 1978, 'Ethical Controversies in Medical Research: The Case of XYY Screening', *Perspectives in Biology and Medicine,* **21**, 196–204.

[17] Kopelman, L. and J. Moskop: 1981, 'The Holistic Health Movement: A Survey and Critique', *The Journal of Medicine and Philosophy,* **6**, 209–235.

[18] MacIntyre, A.: 1967, 'Egoism and Altruism', in P. Edwards (ed.), *The Encyclopedia of Philosophy,* Vol. 2, Macmillan Publishing Co., Inc., and The Free Press, New York and London, pp. 462–466.

[19] Margolis, J.: 1955, 'That All Men Are Created Equal', *The Journal of Philosophy* **52**, 337–346.

[20] Mercer, J. R.: 1973, *Labeling The Mentally Retarded,* University of California Press, Berkeley.

[21] Midgley, M.: 1978, 'The Objection to Systematic Humbug', *Philosophy* **53**, 147–169.

[22] Murphy, J.: 1978, 'Rights and Borderline Cases', in L. M. Kopelman and F. G. Coisman (eds.), *The Rights of Children and Retarded Persons,* Rock Printing Co., Rochester, pp. 6–20 (Reprinted in *Arizona Law Review* 19, 1978, 228–241). Reprinted in this volume, pp. 3–17.

[23] Nozick, R.: 1974, *Anarchy, State, and Utopia,* Basic Books, New York.

[24] Osborne, H.: 1975, 'The Concept of Tragedy', *British Journal of Aesthetics* **15**, 287–293.

[25] Peffer, R.: 1978, 'A Defense of Rights to Well-Being', *Philosophy and Public Affairs* 8, 65–87.

[26] Potter, R. B.: 1977, 'Lebeling the Mentally Retarded: The Just Allocation of Therapy', in S. J. Reiser, A. J. Dyck, and W. J. Curran (eds.), *Ethics in Medicine,* M.I.T. Press, Cambridge and London, pp. 626–631.

[27] Rawls, J: 1971, *A Theory of Justice,* Harvard University Press, Cambridge, Massachusetts.

[28] Singer, P.: 1979, *Practical Ethics,* Cambridge University Press, London.

[29] Spicker, S. F.: 'Person Ascriptions, Profound Disabilities and Our Self-Imposed Duties: A Reply to Loretta Kopelman', in this volume, pp. 87–98.

[30] Williams, B.: 'The Idea of Equality', in P. Laslett and W. G. Runciman (eds.), *Philosophy, Politics and Society,* Second Series, Barnes and Noble, New York.

[31] Williams, M.: 1980, 'Rights, Interests, and Moral Equality', *Environmental Ethics* 2, 149–161.

[32] Wright, B. A.: 1979, 'Atypical Physique and the Appraisal of Persons', in S. F. Spicker (ed.), *Connecticut Medicine* 43, 19–24.

STUART F. SPICKER

PERSON ASCRIPTIONS, PROFOUND DISABILITIES AND OUR SELF–IMPOSED DUTIES: A REPLY TO LORETTA KOPELMAN

Although various authors, especially philosophers, have written extensively on the concept of person – expanding and expounding on the necessary and sufficient conditions which warrant the non-trivial version of the claim that 'Persons ought to be respected' – Professor Kopelman has focused her attention on a more specific problem and on our relation to (what I shall call) 'atypical individuals', though this fine tuning has not made matters any less complex. On the contrary, as R. S. Downie and Elizabeth Telfer acknowledge, just prior to the close of the first chapter in their *Respect for Persons*, and after turning from their consideration of persons to atypical persons, "we are still left with some cases which are difficult to explain in terms of respect for persons." Indeed, Professor Kopelman apparently would agree with their claim that although these cases "present difficulties to any theory of morality, [this] is not really an excuse for failing to make an attempt to account for them" ([2], p. 34). Whereas Downie and Telfer cite the cases of "children, the senile, lunatics and animals" and observe that these cases present difficulties, Kopelman directs her thought over a range of individuals who are differentiated as *mildly, moderately, severely* and *profoundly* retarded. She is, of course, fully aware of the value-laden baggage which such language is compelled to bear; she herself has carefully illustrated that a variety of everyday ascriptions of atypical individuals have built-in limitations. But just so long as general usage and ordinary language engender more social good than ill, on Kopelman's view, such problematic usage may be permitted.[1] With this I find no cause to quarrel.

Moreover, I personally admire Professor Kopelman for directly confronting some very difficult cases and for proffering a few rather interesting reasons for treating retarded individuals with respect – the central theme of this volume. It should of course be noted at the outset that the term 'retarded' serves in fact to explain nothing; at most it is vaguely descriptive and, as Kopelman has adequately demonstrated, troublesomely normative. The term itself does not offer more than a clue as to the kinds of cognition and conation at work in atypical individuals, who have profound mental disabilities and are often perceived as socially burdensome. A detailed account of the spectrum of human atypical cognition and conation properly belongs to the disciplines

L. Kopelman and J. C. Moskop (eds.), Ethics and Mental Retardation, 87–98.

of genetic psychology and genetic epistemology, special variants and extensions of the pioneering work of the late Jean Piaget. For obvious reasons I shall not take up such considerations here.

Rather, as commentator, it is my task to underscore the importance of some of the central arguments proffered by Professor Kopelman and, where I am able, to elaborate on certain specific matters. Although she and I are not in perfect agreement on all crucial points, in general we share in the worries generated by (1) the everyday *language* employed in discussing matters pertinent to atypical individuals, (2) the apparent absence of an adequate description of the proper *objects of respect*, (3) the question of whether or not atypical individuals (who in some cases apparently cannot participate in *reciprocal* relations with persons) can be properly said to warrant an *attitude of respect* on the part of typical persons,[2] (4) the complex relation between an *attitude of respect* and a *principle of respect*, the latter realized, for example, in the *actual* medical treatment of and behavior toward atypical individuals and (5) the problem of what, if any, *right to respect* atypical individuals should enjoy as participants in and members of the human community. I turn then to these five concerns.

Whatever we may take 'persons' to denote it is clear that the term refers to beings thought valuable. Just why persons are valuable is itself not a simple question. Generally speaking, philosophers have tended to agree that a list of features which can serve to describe the independent, generic human self constitutes a good and reasonable beginning. So the literature reveals discussion of features like 'having a rational will', 'being self-conscious', 'acting as an agent', and 'being the source of one's own destiny'; having relatively permanent virtues and vices, like kindness, courage, honesty, cruelty and integrity; and possessing some specific abilities and talents.

A list of such features, dispositions, traits and abilities, is, however lengthy, intrinsically incomplete and misleading. Such person-descriptions, which presuppose that a person is a spatio-temporally identifiable entity, independent in these and other ways from relation to other persons, will continually serve to mislead us in our search for the grounds for treating atypical human individuals with respect. Please note that I have not as yet claimed that atypical individuals, for example, profoundly retarded individuals, are in fact deserving of an *attitude of respect*, nor have I intended to suggest that we should act toward such individuals from an objective standpoint and apply toward them a *principle of respect*. I simply mean to suggest that there is an important connection between our concept of *person* and (1) our participatory human *attitudes* toward and (2) our actual morally justified treatment of a spectrum of atypical individuals.

Leaving the value-laden aspects of language aside for the moment, it is important to give consideration to what, if any, *kinds of objects* are properly said to be deserving of respect. Aside from the commonplace ascription of 'person' to individuals like ourselves, we frequently impute person status to human infants, for example, because of the value we place on the mother-child relationship – its dyadic nature – as well as to protect the integrity of similar social relations ([9], pp. 44–63). But as we know from the plethora of recent literature in bioethics, things become vague when persons make claims on behalf of the person status of fetuses, infants and (pertinent to our concerns) the whole range of retarded individuals, such that these individuals are frequently said to possess certain *rights* [What is at issue here, among other things, is whether or not severely, extremely, or profoundly retarded individuals are persons in any sense, in spite of the tendency of Professor Kopelman to presume and even assert that they are: "many retarded persons deserve respect as persons capable of autonomy . . . " ([7], p. 74). I believe she has not always emphasized the importance of *sharply* distinguishing the two lines of argument which an analysis of agency requires when considering, on the one hand, borderline and mildly retarded individuals and, on the other, extremely and profoundly retarded individuals, though she does occasionally acknowledge "the *varied group* of persons we regard as retarded" ([7], p. 72). There are, to be sure, sufficient resemblances between these two extreme groups, but we must be careful that we do not conflate them. Furthermore, there is also the danger of oscillating between a discussion of such individuals and the broader set of persons in the more unchallenged sense, i.e., people like us.]

Given our concern – what is appropriately ascribable to persons in the strict sense and, say, retarded individuals across the spectrum from mildly retarded to profoundly retarded – John Macmurray's and P. F. Strawson's innovative conceptual schemes turn out to be extremely helpful. In particular, they suggest a novel approach to problems like whether or not it makes sense to speak of our "*respect* for the self of profoundly retarded individuals."

Macmurray views persons (or persons in relation) as irreducibly social; that is, an individual exists when it is in a dynamic relation with the Other. Its being is confirmed, so to speak, in this relating. "'I' exist only as one element in the complex 'You and I'" ([6], p. 169). On this interpretation it follows that a person is a human being, a biologically and genetically endowed individual of a specific sort, but such a human being, completely alone in the world (unless defining himself or herself in terms of his/her relations to other persons), would not be a person.

One corollary which follows from this conceptual scheme is that if the 'I-You' is the unit of the personal, community becomes a special characteristic of being a person.[3] So whereas every human being is an individual, the notion of individual is not the unit of personal existence. Hence only in the personal relations of persons is personal existence constituted. On this interpretation there can be no person in strict isolation from all other persons. A person comes into existence when there are at least two selves in actual or imagined relation or communication ([8], p. 12).

In Strawson's system, we begin with "the primitiveness of the concept of a person."[4] That is, 'person' denotes a type of entity "such that *both* predicates ascribing states of consciousness *and* predicates ascribing corporeal characteristics, a physical situation &c. are equally applicable to a single individual of that single type" ([10], pp. 97–98). Hence, breaking with the long-standing Cartesian tradition, we now abandon our conception of a person as some compound of two kinds of subjects – a subject of experiences and a subject of corporeal or material attributes. It is very important to appreciate these novel conceptual schemes wherein the concept of person is logically prior to that of an individual consciousness. Strawson, we may recall, employs the notion of P-Predicates in all descriptions of persons (which includes both predicates which ascribe states of mind or consciousness *and* predicates which ascribe corporeal or material characteristics to these individuals). [M-Predicates are those predicates which apply to all material bodies ([10], pp. 100–101). Although persons and material objects may be described by the same predicates, this does not create a serious problem within Strawson's scheme of predication.] For our purposes, we are simply concerned with a very special class of P-Predicates, what I shall call Personal-Relational-Predicates (which I shall designate 'P-R-Predicates'). So predicates like 'the-ability-to-be-loved', 'the capacity-to-call-for-compassion' and 'the-ability-to-reciprocate-interpersonally-within-the-human-community' are P-R-Predicates and constitute, to the degree that such a list is more and more complete, the meaning of 'person', *in addition to* the more conventional list of P-Predicates proffered by the idealistic philosophical tradition, like 'is intelligent', 'is self conscious' or 'has a rational will'.

Given that both P-Predicates and P-R-Predicates are necessary to any adequate description of persons, what is the status of retarded or profoundly retarded individuals? There is little doubt that they are, like the rest of us, human beings with a similar genetic endowment.[5] But this latter notion is essentially biogenetic and of virtually little import for those who wish to argue that severely and profoundly retarded individuals are persons deserving of our respect, esteem and honor – that is, that *we* should treat *them* from *an attitude of respect* as a *logically appropriate* attitude. Notwithstanding Professor Kopelman's effort to warn us of the way language betrays our true beliefs about how we value retarded individuals, we could be easily led to another equally troublesome position: to presume all too quickly that profoundly retarded individuals, and perhaps even others less profoundly handicapped, are *entitled* to our attitude of respect and have a *right* to respectful treatment. My point is that Professor Kopelman has not as yet made her case, in spite of her four-fold senses of 'respect'. For to make her case – that we ought to respect profoundly retarded persons – it is not very helpful to appeal to her first sense – esteem, respecting the work or products of persons; nor is it to the point to appeal to her third sense of respect – our regard for the circumstance of the retarded and the fact, say, that a significant genetic anomaly has led to a condition beyond the individual's control.

Professor Kopelman, aware of those senses which miss the mark, focuses on her second sense of respect – the ability or capacity of retarded individuals to act as the source of their own destiny and to "work things out for themselves" ([7], p. 72). At one point she suggests that respect in the sense of esteem is appropriately directed to retarded individuals since respect can be used in "the sense of honor or esteem." But I do not see how she can justifiably claim that such conclusions are sound; for the case of Brother William is not really a case of a community's *attitude* of respect, though it may well be an example of a community *acting from a principle of respect,* which is an entirely different matter. For Brother William is not an appropriate object of respect in the sense of being the object of our attitude of respect. I say this for the following reasons: For an individual to be a proper object of respect that individual must be capable of rule-following, self-control and possess some level of knowledge which warrants an *attitude of respect;* such an individual must be capable of sustaining our trust and, most importantly, be able to enter into a *reciprocal relationship* with persons (i.e., be described by P-R-Predicates as well) who are able to enter into the entire range of our human participatory attitudes. In short, a proper object of respect must also be able to be the *source* of a similar attitude. If retarded individuals can be shown to be capable of such dispositions, knowledge, and rule following (and this is an empirical matter I must leave to the experts – psychologists, physicians and others –)[6] then they can be the objects of an *attitude of respect.*

At this juncture, then, we should at least sort out the *kinds of attitudes* which we can properly be said to adopt toward persons in the strict sense – those essentially like ourselves – and other human individuals who are more and more unlike our relational selves in not inessential ways. Once again Strawson offers us a very fruitful distinction.

In 'Freedom and Resentment' (1962) Strawson distinguished "reactive attitudes and feelings" which emerge in our daily involvement and participation with other persons, from "objective attitudes" ([11], p. 79 *et passim*). These two sets of attitudes, though they are not altogether exclusive of each other, are "profoundly *opposed* to each other." The participatory attitudes insinuate our ordinary interpersonal relationships of the 'I-You' mode, from the most intimate to the most casual. But as the Other becomes less and less an Agent we find that the reactive attitudes, *which entail a reciprocity on behalf of the Other*, become less and less appropriate, and we find ourselves more and more operating from various objective attitudes.[7] These objective attitudes are based on *objective principles*. For those retarded human individuals who are only capable of reacting from the standpoint of the non-moral attitudes (e.g., affection, frustration, repulsion, attraction) *we* have to respond at times from the standpoint of certain objective attitudes. The ordinary moral attitudes (e.g., respect, dignity, approbation) which we tend to direct toward other persons like ourselves become inappropriate when directed to those individuals who have virtually no capacity to reciprocate toward us from the standpoint of these moral attitudes. In short, we tend to (and ought to) view profoundly retarded human individuals in a very different light from the light in which we normally view other persons like ourselves. In my judgment this is not to advocate social prejudice nor bias, but is totally appropriate. The "morally incapacitated," if you will permit the term, may well be the appropriate objects of our *objective attitude of respect,* but they cannot be the appropriate objects of our positive participatory attitudes of respect, esteem or honor. The objective attitude is appropriately present when we undertake the treatment and care of severely and profoundly retarded individuals; the objective attitude is equally at work when we formally adopt a *principle of duty* toward other human individuals who cannot reciprocate *via* the participatory moral attitudes. Most importantly, although such objective moral attitudes as, for example, acting from duty alone, may be emotionally laden, they are in some sense beyond the range of reactive feelings and attitudes which usually come into play in our involvement and participation with other persons in our everyday interpersonal human relationships. Not only the moral attitudes of respect, esteem, honor or approbation require reciprocity but so do *agape,* gratitude and anger. If we carefully consider the nature of our relations with severely and profoundly retarded individuals, we find that we can at most *pretend* to relate by way of the moral-participatory, reactive attitudes. Although Professor Kopelman mentions the fact that a "lack of reciprocity" becomes a problem when respect is extended to the profoundly retarded ([7], p. 78), she does not in my judgment adequately attend to the importance of such interpersonal reciprocity in drawing her conclusions.[8]

Now suppose there exists a profoundly retarded human individual who is devoid of the specific capacities (moral attitudes) I have mentioned, and therefore can make no claim on our *attitude* of respect. What follows from this? Has that individual's entire personhood been forfeited under this description? I think not. For it is here that we can make a valid claim that such an individual still warrants respectful treatment under a *principle,* not an attitude, of respect. For with regard to such a profoundly retarded individual we are perfectly consistent in denying that he or she deserves or has earned our attitude of respect, although he or she may be treated from the principle of respect as a self-imposed duty. Again, such individuals may be objects of our concern, kindness and consideration, though I seriously doubt whether they could be the objects of respect, awe and admiration. Such relations do indeed reflect a moral point of view toward such atypical individuals, but it does not follow that having adopted a principle of respect for such individuals that we should, or could, logically speaking, adopt an attitude of respect toward them. An attitude of respect is intrinsically related to a full sense of personhood and it is the latter that is being challenged in the case of the profoundly retarded individual.[9] (Furthermore, adopting a principle of respect – in the objective attitude – toward such individuals is quite consistent with treating such individuals as less than autonomous.)

But what, if anything, justifies the adoption of a principle of respect toward those individuals with profound mental disabilities? The answer to this question lies in our ability to adopt an objective attitude and *impose on ourselves a principle of duty and obligation toward others,* which duty is not symmetrically related to anyone's claim to a right or 'to have a right to x'. That is, whereas my *right* to respect and respectful treatment *entails* that others have obligations toward me, it does not follow that our self-imposed obligations entail that profoundly retarded individuals, or any individual, has, for example, a *right* to respect and respectful treatment. Note also that we often speak of "earning" the respect of others; here it would be difficult indeed for our *severely* or *profoundly* retarded individual to lay claim to having *earned* anything in particular.

Taking the notion of respect seriously, then, will not permit the profoundly retarded individual to lay claim to respect, except as *we will* to adopt a self-imposed obligation toward such individuals, who by their very presence are members of the human community. From our more advantageous position (or perhaps from behind some Rawlsian veil of ignorance) we can *bestow* rights on the retarded members of the community. From this *derived sense* of respect we can then properly be said to 'respect the rights of the retarded'. But it must be kept in mind that this is precisely a derivative sense of rights. The failure to distinguish between *respect as an attitude* toward persons and various derivative usages has led to all kinds of confusion, and permits us to continue to pretend that dependent and not very autonomous individuals are more free and capable than they in fact are. This too can lead to abuse and the inconsiderate treatment of such individuals. Hence we must be alert to the derivative senses of 'respect' when the term is employed to signal the power of a disease (Kopelman's fourth sense), the danger of high voltage wires, or one's enemy's skills. All of these nonpersonal objects are said to be capable of being "treated with respect". But none of these illustrates the strict notion of respect – *respect for persons.* Furthermore, such derivative senses of 'respect' as "I respect the power of tuberculosis or this dynamite" clearly indicate that we are using a person predicate (e.g., respect for) when we clearly should restrict ourselves to M-Predicates in these cases, since dynamite and bacteria are not persons but forms of corporeal existence. What shall we conclude from all this?

Whereas some philosophers seek to establish a defensible respect-for-persons theory, I suggest we seek to establish a defensible *theory of obligation* which will be independent of both a respect-for-persons theory [1] and perhaps all rights theories. From a principle of respect, which is simply an element within a fully worked out theory of obligation, one can *act* respectfully toward the disenfranchised, the severely disabled, and the very dependent. This may even prove a sounder basis for all medical practice (but I shall not undertake the arguments to show that here). So when Professor Kopelman says that "we rarely give retarded persons respect" ([7], p. 65) she has in fact made an astutely important observation. But it may be more fruitful to wonder at the subtle reasons for this social fact, rather than offer an indefensible line of argument: that we ought to maintain an attitude of respect toward such individuals. The notion of 'respect' is, then, to say the least, ambiguous.

Treating persons from an attitude of respect and consideration is the hallmark of a civilized community. A theory of duty and communal obligation could conceivably free us from the problems posed by severely and profoundly retarded individuals which Professor Kopelman foresees. No inhumane treatment on this view would ever be justified within the sphere of communal interaction with regard to the handicapped and otherwise dysfunctional individuals. The careful articulation of a doctrine of obligation in codified civil and criminal law can serve as adequate protection for those individuals whom others may neglect or even injure. It seems to me that we might finally honor dependency not by chastizing ourselves for acting paternalistically, but by treating the dependent from the principle of duty, i.e., with respect – acknowledging that some of us have some advantages which require less protection – thereby ennobling ourselves by the deeds we do from duty alone.

In conclusion, membership in the human community is properly grounded in our ability to adopt various self-imposed duties toward others. And here I am reminded of an obscure passage in Kant's *Grundlegung* in which he mentions those actions, like helping others, which are motivated from sympathy and inclination (i.e., participatory attitudes and motives) but which have no genuinely moral worth and which appropriately deserve support, acknowledgement, and encouragement, but not esteem or respect. For actions to be of genuine moral worth they must be done from duty alone. Ironically, perhaps, actions which are motivated from duty alone, and which therefore have genuine moral worth, are precisely those actions on the part of an Agent which are carried out in the objective attitude. I turn to Kant's illustration of an action deserving moral worth: "Suppose then that the mind of this friend of man were overclouded by sorrows of his own which extinguished all sympathy with the fate of others, but that he still had power to help those in distress, though no longer stirred by the need of others because sufficiently occupied with his own; and suppose that, when no longer moved by any inclination, he tears himself out of this deadly insensibility and does the action without any inclination, for the sake of duty alone; then for the first time his action has its genuine moral worth" ([5], p. 66).[10]

How does this Kantian illustration apply to what I have argued thus far? Persons like ourselves, who in the ordinary course of events care for, support, respect and hold other persons in esteem and honor would, in acting in these ways toward one another, have acted from motives of inclination. But all the more would our actions be of genuine moral worth if, motivated to act from duty alone, we would adopt the objective attitude in our treatment and relations with other persons and all profoundly retarded human individuals and perhaps even those less afflicted and of greater capacities. What is required of us are self-imposed duties and self-generated laws, which permit the bestowing of rights on those less fortunate than ourselves. Those individuals beyond the reach of ordinary interpersonal relations and the moral attitudes should at least be treated from the standpoint of the objective attitude. (It should be noted that we may also adopt this attitude with respect to typical persons who, on occasion, due to their antisocial behavior and conduct warrant such a standpoint. Thus we may not infrequently adopt the objective attitude with respect to "normal" persons.) But what we have to face, however unhappily, is the fact that some human individuals are truly incapacitated in some or virtually all respects, and hence cannot warrant or require that persons treat them from the standpoint of the participatory moral attitudes. (Even the language of philosophers has, on occasion, reflected less than kind perjorative implications noted in such phrases as "moral idiots" ([11], p. 86) – and not so many years ago at that.)

Retarded individuals, although they cannot command our respect in the primary attitudinal sense, can be the proper recipients of our self-imposed obligations. Notwithstanding the fact that we, on occasion, must suspend our personal moral attitudes toward certain human individuals, we are also so constituted that we can purposively act on their behalf and *for them* from the strict motive of duty alone. In so doing our actions may be judged, perhaps for a brief moment, of genuine moral worth. Ironically, those individuals who may be wholly lacking in moral sense may actually afford the rest of us those rare opportunities to act from the motive of duty alone, and thereby to earn mankind's respect, esteem and honor, for which, ever since Spinoza at least, it is said that we all strive.

University of Connecticut
School of Medicine
Farmington, Connecticut

NOTES

[1] The language employed by researchers in England in 1960 is most revealing and serves as an example of the linguistic stigmatization noted by Professor Kopelman:

"Opinions differ concerning the upper end of the range of intelligence in mongols. Some would maintain that none are above imbecile level; others that a significant proportion are only feeble-minded" ([3], p. 565).

2 In my previous remarks I have, as you may have noticed, refrained from prejudging whether or not individuals with congenital disabilities – the result perhaps of powerful genetic determinants like Down's Syndrome – are or are not persons in some strict sense. To be carefree when employing the notion of 'person' in this context is to be philosophically irresponsible or, more simply put, to beg one of the central questions which we shall have to face.

3 Kopelman also seems to follow Macmurray and Strawson here, when she remarks that "we stand in certain relationship to each other as human beings" ([7], p. 79) and when she presupposes a social context for "mutual respect" ([7], p. 74).

4 It would be very interesting to pursue the similarity in standpoint between P. F. Strawson's position and John Macmurray's earlier Gifford Lectures of 1953 (See [8]) and 1954 (see [9]). They are in fundamental agreement that the concept of person is logically prior to concepts of 'Self', 'Mind', 'Consciousness', etc.

5 If there is a good case for pointing to our rationality and moral personality as the bases of the significant moral distinction between persons and non-persons, then this should, perhaps, be maintained in spite of marginal instances, such as the severely mentally retarded. As David C. Hicks has remarked: "Why not have the courage to accept the logical implication of this position, viz. that there doubtless are many creatures who, although born of the human race, do not qualify as persons" ([4], p. 347).

6 Professor Kopelman seems to agree that the empirical states of affairs are not always easily determined, as in "estimating the extent to which they [retarded adults] are capable of making decisions about themselves, for themselves" ([7], p. 74). Note also her use of 'human individuals' ([7], p. 67). Is this to avoid, as I have done, presuming that retarded individuals are persons?

7 The importance of P-R-Predicates lies, in part, in the fact that they are frequently employed in referring to the participatory *reactive* attitudes of our everyday human interactions. In those interactions between persons and severely retarded individuals, P-R-Predicates are not appropriately employed, as *reciprocity* is sorely lacking. In such contexts persons tend to adopt the *objective attitude* and continue to employ what Strawson calls P-Predicates. Given Strawson's distinction between participatory and objective attitudes, I would imagine that he would accept the additional sub-category of P-R-Predicates, since the negation of some P-Predicates alone would not serve very adequately to describe those individuals, for example, who are, in his words, "moral idiots" ([11], p. 86).

8 One place where such distinctions are glossed over may be found in her remarks about "mutual respect." The "mutuality" is not well grounded in reciprocity, and her discussion tends to be focused on persons in the strict sense ([7], p. 74).

9 Although Professor Kopelman properly warns us of a danger that accompanies the tendency to "reinforce an association of retarded persons with children . . . thereby increasing the risk of ignoring their need for respect as persons who set their own goals and make their own plans" ([7], p. 70), I do not think it follows that severely retarded individuals in general are persons, or persons deserving of respect. The comparison of the very elderly, as well as retarded individuals, with children (and the description of their

behavior as "childish") is, to be sure, not only invidious but also in total disregard of the empirical, genetic psychological facts. Retarded individuals, are not just "slower" to develop then "normal" children. Rather, they develop in a very specific set of ways qualitatively quite unlike those of "normal" children.

[10] The original text [S. 398] reads:

"Gesetzt also, das Gemüt jenes Menschenfreundes wäre vom eigenen Gram umwölkt, der alle Teilnehmung an anderer Schicksal auslöscht, er hätte immer noch Vermögen, andern Notleidenden wohlzutun, aber fremde Not rührte ihn nicht, weil er mit seiner eigenen genug beschäftigt ist, und nun, da keine Neigung ihn mehr dazu anreizt, risse er sich doch aus dieser tödlichen Unempfindlichkeit heraus und täte die Handlung ohne alle Neigung, lediglich aus Pflicht, alsdann hat sie allererst ihren echten moralischen Wert" ([5], pp. 37–38).

BIBLIOGRAPHY

[1] Cranor, C.: 1975, 'Toward a Theory of Respect for Persons', *American Philosophical Quarterly* 12(4), 309–319.

[2] Downie, R. S. and Telfer, E.: 1969, *Respect for Persons,* George Allen and Unwin Ltd., London.

[3] Dunsdon, M. I., Carter, D. O., and Huntley, R. M. C.: 1960, 'Upper End of Range of Intelligence in Mongolism', *The Lancet* (March 12), 565–568.

[4] Hicks, D. C.: 1971, 'Respect for Persons and Respect for Living Things', *Philosophy* 46, 346–348.

[5] Kant, I.: 1959, *Grundlegung zur Metaphysik der Sitten* (1786) (dritte auflage) Theodor Valentiner (heraus.) Reclam-Verlag, Stuttgart. For the English translation see 1956, *Groundwork of the Metaphysic of Morals* (transl. H. J. Paton), Harper & Row, New York.

[6] Kjaergaard, Astrid: 1970, 'Action and the Person: Macmurray's *The Form of the Personal'* (Review Article), *Inquiry* 13, 160–198.

[7] Kopelman, L.: 1984, 'Respect and the Retarded: Issues of Valuing and Labeling', in this volume, pp. 65–85.

[8] Macmurray, J.: 1957, *The Self as Agent,* Faber and Faber Ltd., London.

[9] Macmurray, J.: 1961, *Persons in Relation,* Faber and Faber Ltd., London.

[10] Strawson, P. F.: 1959, *Individuals: An Essay in Descriptive Metaphysics,* Methuen & Co., London.

[11] Strawson, P. F.: 1968, 'Freedom and Resentment' in P. F. Strawson (ed.) *Studies in the Philosophy of Thought and Action,* Oxford University Press, London, pp. 71–96.

LAURENCE B. McCULLOUGH

THE WORLD GAINED AND THE WORLD LOST: LABELING THE MENTALLY RETARDED

The subject of labeling the mentally retarded is fraught with controversy. There are, for example a host of empirical questions. Does the label 'mental retardation' affect how others think about and act toward retarded individuals or the retarded as a group? How do the retarded, especially the mildly retarded, respond to being so labeled? Do they find this designation burdensome? a stigma? or a matter of indifference? These and many other questions have been subjected to intensive empirical study in the past several decades [18, 29, 31]. They will not, however, be my chief concern. Instead, I will inquire into the more strictly conceptual and ethical dimensions of labeling.

I view a label primarily as a verbal or linguistic device that, in its most fundamental terms, creates and sustains a world. It is a performative utterance [3, 4]. What one gains for the mentally retarded through a label is the basis for an entitlement to special treatment aimed at overcoming or at least ameliorating one's handicap [33, 30]. Hence the first part of the title of this paper. At the same time, however, one also loses a world for the retarded, the world of the rest of us who are, as it were, "normal." In particular, the label of mental retardation, like other labels such as mental illness, seems linked to the loss of certain prerogatives, most notably a standing of full moral worth as a bearer of rights and responsibilities and thus of fair and equitable opportunity to enjoy life to the same extent as those who do not bear the label. Hence the second part of the title of this paper. But must the first always be purchased at the price of the second?

I want to argue that the two worlds are not always logically connected. That is, only on a restricted understanding of the label is it necessary that the world gained through the label of mental retardation must be purchased at the price of a world of diminished moral worth or diminished rights and responsibilities. By contrast, the common understanding of the label, based on the definition of mental retardation offered by the American Association on Mental Deficiency [17] suggests a loss of moral worth which, I hope to show, cannot be justified. The root problem, in my view, is a conceptual muddle surrounding the definition of mental retardation. I will, therefore, try in this paper to bring clarity to the definition in a way which also brings clarity to the ethical dimensions of the problem of labeling. I begin with an

L. Kopelman and J. C. Moskop (eds.), Ethics and Mental Retardation, 99–118.

analysis of some considerations of justice in the treatment of the retarded and of what it is to be handicapped, elements which will come to figure prominently in my argument.

I. JUSTICE AND THE RETARDED

How does the label of mental retardation gain for its bearers a world in which special treatment is an obligation? The root of this obligation, in my view, is in an entitlement to sustained opportunities for development of one's capacities and life projects. By 'sustained' here I mean more than a 'minimum decent standard' of care [15] and less than a maximum standard. The problem with a minimum standard is that it ensures only that opportunity for human achievement is initially open to one, not that it will *remain* open. The problem with a maximum standard, as others have recognized in such areas as health care [6], is that it is not achievable unless a society is able to generate unlimited resources. Thus, 'sustained' is a necessary modifier because it reminds us that the goal of special treatment is to keep opportunities for the mentally retarded open to them for the full course of their lifetimes, just as we ought to do for people in general.

Making sense of such an entitlement involves basic questions in ethical theory, and so it cannot be undertaken here in full detail. Let me, however, suggest two ways of proceeding. The first approach is a rather traditional one, joining metaphysical anthropology, ethics, and political theory together to establish a view of human nature and of just social orders in which human beings would be allowed, even encouraged, to achieve the fullest expressions of human nature. This approach is based on natural or human rights and inspired by John Locke. Though they cannot be absolute, human rights may serve as morally significant constraints on public policy. As W. T. Blackstone has put it:

> Those who have spoken of human rights as universal and inalienable have not intended to assert that the actual exercise of these rights on a given substantive human right may not properly be denied or overriden if the force of other morally relevant considerations is stronger in a given situation. What we can *never* do is rule out a human right as a morally relevant consideration ([5], pp. 627–628).

That is, the *exercise* of natural rights is properly limited. Hence, such an approach can support a right to "sustained," but not "maximum" opportunity. Any policy on this view, however, that ruled out or denied human rights altogether would be unjustifiable.

A second approach to grounding a right to special treatment for the retarded makes use of John Rawls' theory of justice [34]. Rawls uses a hypothetical choice situation, which he calls the 'original position', to establish basic principles of justice. Since the individuals in the original position do not know their station in life, characteristics, capacities, etc., they would, I suggest, want to ensure that the fullest possible range of opportunities remained open for a sufficient period of time, so that they could be secured and exploited. Since some individuals, such as the retarded, will need assistance in realizing their opportunities, there appears to be a prima facie rationale in Rawls' theory for providing special programs for the retarded. This conclusion gains further support from the requirement of fair equality of opportunity in Rawls's second principle of justice, because in the absence of special treatment, the retarded will probably suffer a loss of opportunities for human development compared to those of "normal people." Yet the history of our treatment of the retarded in American society has produced just such losses, which we have, belatedly, come to view as unjust. Modification of the Rawlsian account to include the notion of "sustained opportunity" allows us clearly to see why such policies as mainstreaming in education may be justified if they enhance the opportunities of retarded persons.[1]

I do not mean to imply that Rawls' theory of justice or a revision of it can be applied in an altogether routine manner to the full range of public policy questions regarding special services for the retarded. As Norman Daniels [10] has pointed out, the Rawlsian approach clearly seems to justify special treatment in such areas as education and job training. In the area of health care, however, Daniels raises serious questions about the attempts of those who, like Ronald Green [16], have argued that health care should be treated as one among the primary social goods and thus subject to Rawls' concept of distributive justice. This dispute, however, does not undercut my main thesis that special treatment of the mentally retarded is an obligatory matter in justice (for Locke or Rawls), since it is about what *types* of services will count as justified special treatment rather than whether special treatment is itself justified.

So much for questions of special treatment. My main task in this paper is to explore the following question: must special treatment, the world gained by the label of mental retardation, always be purchased at the price of diminished moral status? The history of our social and legal response to the retarded has been to strip them of rights and responsibilities across a broad spectrum – educational and job opportunities, housing in the community, voting, marriage, reproduction, etc. – simply because they are more than two standard deviations below the mean of intelligence. I shall argue that this sweeping loss of rights and responsibilities is not legitimated by the label of mental retardation, properly understood. I begin this argument with a reflection on what it is to be handicapped, in terms of which concepts of task-specific competency and task-specific incompetency are developed. These concepts are then applied to the standard definition of mental retardation, that of the AAMD, in a way that brings clarity to that definition. The result is to show that the loss of rights and responsibilities legitimated by the label of mental retardation is an individual matter, one that must be argued on a case-by-case basis and not solely on the basis of membership in the class of those (accurately) labeled mentally retarded.

II. BEING HANDICAPPED

When one validly assigns the label of mental retardation to someone, one is thereby signifying that that individual is handicapped. This would seem, at first, to be a straightforward word. One is handicapped if one is disadvantaged by a disability of one sort or another: physical, mental, emotional, motivational, social, and so on. Because particular human achievements are more difficult for the handicapped person to undertake, he or she may not secure them or even get the chance to secure them for him or herself. The reasons for this, as we shall see, include our attitudes toward those handicaps and the persons who have them as well as our willingness and ability to make available resources that would minimize or perhaps eliminate the handicap in question.

Handicaps usually happen to people without their having done anything to cause them. One is simply born blind or deaf or, the phenomenon we are examining in this conference, mentally retarded. We have a name for such occurrences: we call them 'acts of God'. By this we mean that the handicap is fortuitous, undeserved. The person with the handicap, it is reasonable to say, did nothing to cause the handicap to occur and neither did anyone else. We say, in a common phrase, 'things happen like that'.

There is another sense of handicapped, though, that is quite different. This is the sense in which an advantage or disadvantage is imposed on someone by someone else. Thus, a golfer who regularly plays in the eighties is assigned a handicap when in competition play. He or she thus gains an advantage over the golfer who regularly plays in the seventies. The first golfer's chances of winning are increased by the handicap assigned. It may well be that this is considered a fair advantage for the purposes of tournament play but it is an artificially imposed advantage nonetheless. The word handicap also signifies that disadvantages can be imposed for certain purposes. Thus, before many horse races, jockeys must weigh in with their saddles and be assigned additional weight as a handicap so that all the horses carry equal weight and no horse and rider gains unfair advantage over another. These handicaps are not acts of God but deliberate, self-conscious choices whose consequences are artificially (in Aristotle's sense) imposed on those affected.

I suggest that when we speak of fellow human beings as being handicapped in the first sense – i.e., they simply happen to suffer a particular disadvantage that others do not – we in very subtle ways also impose a handicap in the second sense. That is, our attitudes, value judgments and practices often place the handicapped at additional disadvantage. That is, saying that someone is handicapped and signifying that quality with a label often places him or her at a double disadvantage: he or she happens to be disadvantaged and our attitudes and value judgments make him or her so. Since these attitudes and judgments are the foundation of our behavior and social practices regarding the handicapped, the disadvantages thus imposed on the handicapped in virtue of having been assigned a label appropriately (from the stance of the first sense of being handicapped) take on ethical significance. One may be unjustifiably denied opportunities for the pursuit of human achievement.

We might call these two senses the intrinsic and the extrinsic value dimensions of labels for handicapping conditions. The intrinsic value dimension of a handicap bears on the diminishment that the handicap *per se* causes. That is, this intrinsic sense concerns the value dimensions of a handicap without reference to how others respond to the label and thereby evaluate its significance. Thus, if I am born profoundly retarded there are certain human experiences closed to me, no matter what others may think and do. I will not be able to master the philosophy of Immanuel Kant or the intricacies of modal logic.

On the other hand, how others evaluate a label that implies a handicap may also arbitrarily close off experiences for those labeled as handicapped. Who would have thought twenty years ago that the mentally retarded, for example, could attend neighborhood schools or compete in athletic events? Doesn't one have to be of "normal" intelligence to do these things? How one answers questions about handicaps like mental retardation will, as we now know, have a profound impact on the quality of life enjoyed by the handicapped. That is, how we respond to the behavioral manifestations of mental retardation handicaps as well as "intrinsic dysfunction." This is the extrinsic sense of being handicapped.

Already we can see how our ordinary use of the language of being handicapped embraces complex value judgments with important implications for how we treat those with handicaps. There is still another implicit value judgment imbedded in our language about handicapping conditions. Both senses of the label, 'handicapped', just analyzed can sometimes be taken to imply that being handicapped is a fixed state. One is or is not handicapped and one's disability is a state of one's existence not open to change or improvement – except, of course, those handicaps recognized as temporary, e.g., a broken arm. This assumption leads to a conclusion with significant implications. If one's disability is a fixed state, not open to change, then that's that. And – here is the ethical and policy punch in that conclusion – there is relatively little we can do to make a difference, so let's put our resources elsewhere, where they will presumably do more good.

I suggest that such a view of characteristics basic to human beings cannot be sustained. None of us who are "normal" take ourselves to be fixed in our abilities. If we are not good swimmers, for example, we can exercise and train and improve our skills. It is a particularly vicious, though very subtle form of prejudice that we do not treat handicaps in the same way. Consider Down's syndrome, for example. It is only recently that research has been undertaken to improve the outcome for babies born with this handicap [28]. The possibility of doing that research, I suggest, occurred only recently, as we began to change our minds and adopt a more process-oriented view of the human condition.

When we adopt such a view of handicapping conditions we should recognize that they are not fixed in character, they are not the same for each individual characterized by them. Thus, defects in adaptive behavior attending mental retardation will vary from person to person and in the care of each individual person may be open to different levels of amelioration. Taking the process view of disabilities, however, should not lead us to think that a disabled person's potential has no upper limit. As a consequence, we may set standards that turn out to be unachievable and in some cases destructive of the very potential we seek to develop. This problem is not peculiar to the disabled; it applies to all of us. Nevertheless, abandoning the assumption that disabilities and handicaps have fixed outcomes should help gain us a world in which there will be increased opportunities for the disabled to develop their potential and an obligation on our part to provide those opportunities. We will not know in advance how great any individual's potential will be. We will learn as we begin to press against those limits. Knowing when those limits are near and avoiding promises that could be enormously destructive are special skills we will all need to learn.

Handicaps and disabilities do, of course, impose limits on those they affect, but a failure on our part to understand the flexible character of those limits can impose still another handicap. That is, the intrinsic sense of the label of mental retardation, for example, when wedded to a fixed state view may itself result in an unjustifiable loss of opportunity for experiences which would permit one to develop whatever intellectual capacities one might possess. This loss of a world would occur independently of our attitudes, i.e., independently of the extrinsic sense of the label.

The philosophical task of analyzing value judgments which are often hidden from us because we have not thought carefully about them, has disclosed the need to distinguish the intrinsic from the extrinsic value dimensions of labeling the disabled or handicapped. Reasonable or justifiable attitudes toward the handicapped will avoid ethical judgments that arbitrarily restrict the handicapped person's opportunity to develop his or her potential. Moreover, such value judgments will focus on avoiding the major pitfalls of taking a process view, viz., setting expectations too high. In addition, we do not want to deny to the handicapped the opportunity to be themselves and thus different from the rest of us. We should be interested in normalization and habilitation only when those goals embrace the value of diversity and toleration of differences [20].

In summary, then, we may say so far that labeling the mentally retarded should satisfy the following conditions:

(1) recognition of the distinction between the intrinsic and extrinsic value dimensions of being handicapped or disabled;

(2) rejection of a fixed state view in favor of a process view of being handicapped or disabled;

(3) respect for diversity and difference in both the intrinsic and extrinsic value dimensions; and

(4) avoiding the establishment of destructive expectations for realization of the potential of the handicapped or disabled person.

In the balance of the paper I want to explore still further the first two of these considerations.

III. LABELING AND COMPETENCE

We have noted two basic problems with labeling the retarded. The first is that a fixed state view of retardation in the intrinsic sense of the label causes those so labeled to lose a world. And, it would seem, the extrinsic sense of the label also causes the retarded to lose a world. Since in both cases of loss fundamental ethical questions are raised, they are worthy of more attention. Resolving those questions turns first on how we should understand the notion of competence and second on the bearing of this notion on how we should understand the definition of mental retardation.

The label of mental retardation is tied to the loss of worlds because it implies that one is incompetent: one is thought to suffer a significant diminishment of the intellectual or behavioral capacities that mark the "normal." These capacities are tied directly to moral status as its necessary conditions. This is especially the case with rights and responsibilities.

Consider the following example. Suppose that I am blind, yet claim the right to operate a motor vehicle. Now the right to hold an operator's license requires certain knowledge and skills, among them knowledge of one's state's motor vehicle laws, ability to pass a written test on those laws, and ability to satisfy the license grantor that one can safely operate a motor vehicle. The first two criteria, I have no doubt, the blind could meet with a modicum of assistance, e.g., reading machines and the like. But the third condition, the capacity to operate a motor vehicle safely, the blind person cannot satisfy with any mode of assistance presently available. To insist, nonetheless, on a right to equal access to an operator's license is not justifiable – a skill or capacity required as a necessary condition for the exercise of the right in question is absent and cannot be ameliorated by special treatment. In the face of such an absence of competence, the grounds for the right vanish.

Now consider a much more difficult topic, the sterilization of the mentally retarded. The abuse of this medical technique in the case of the mentally retarded is, in many cases, a moral outrage. I reside in a state where what appears to have been the routine sterilization of the institutionalized mentally infirm took place up until recent times without the subject's consent or that of his or her parents but with the sanction of law, a sanction that – to the best of my knowledge – has not been rescinded.

Even in the face of such a history, however, it is important to ask when and whether the sterilization of the mentally retarded is justified, if only because our emerging ability to adopt techniques of reversible sterilization will press this issue on us again with a renewed vigor [24]. In a recent issue of *The Lancet* the Working Group in Current Medical/Ethical Problems in Great Britain offered a set of recommendations concerning sterilization of the mentally handicapped. In a well-reasoned report this interdisciplinary group recommended that, if sterilization is being used to prevent reproduction of children who might not be adequately cared for, in the case of both women and men, it can best be justified only in connection with parental inadequacy.

There may be several reasons favoring sterilization of an individual patient, but in most cases sterilization should only be undertaken where there is an inability to cope with the demands of parenthood ([19], p. 686).

Notice here the narrow scope of the recommendation. Eugenic considerations have been eliminated, and so cannot be legitimated by the label of mental retardation, a considerable gain over previous views, e.g., those expressed by Mr. Justice Holmes in *Buck v. Bell*. Also, this view rejects the position that mental retardation should diminish rights and responsibilities broadly. Only those competencies related to being a parent are involved. That is, the right to be a parent implies, as its necessary condition, the capacity to meet the demands of parenthood. If one is unable, for reasons beyond one's control, to meet the special responsibilities of a parent, then one cannot reasonably assume them – especially when the welfare of an innocent third party is at stake. In sum, as one's capacities requisite for the exercise of rights and responsibilities diminish, so too do one's rights and responsibilities. And, where that diminishment cannot be made up by the assistance of others, then it is justifiable to circumscribe the rights and responsibilities of the handicapped – just as we should *for any person,* handicapped or not, who fails to fulfill the abilities requisite for the right or responsibility in question. That is, the ethical principle of diminished competence applies to any person, and thus also to the handicapped, when there are reasonable grounds for believing their capacity is diminished.[2]

The ethical principle I have proposed owes a good deal to Robert Cooke, a pediatrician, whose work on the ethical dimension of mental retardation is widely known. Cooke calls for a "doctrine of diminished responsibility" in the area of responsibility of the mentally retarded to be found guilty of a crime.

This doctrine of diminished responsibility states that the defendent is neither wholly responsible nor wholly irresponsible; it calls for recognition of an allowance for varying degrees of mental impairment [9].

That is, mental retardation, according to Cooke, is properly understood as sometimes impairing the capacity to distinguish right from wrong, to a different extent in each individual thus affected. This capacity is requisite to assume public (and criminal) responsibility for one's actions that affect others. When that capacity is reduced and cannot be compensated for by special treatment, so is one's responsibility – in some cases to the point where in individual cases no criminal culpability should be assigned.

Notice what emerges from this analysis of the rights of the handicapped and disabled. It is that one's rights and responsibilities diminish only insofar as one's capacities requisite as necessary conditions for the exercise of those rights and responsibilities diminish and cannot be ameliorated by special treatment. Clearly, though, many people suffer such diminishments even though we would not ordinarily call them handicapped – consider, for example, pleas of temporary insanity or alcoholism in criminal trials. Thus, we should accord the handicapped, and the mentally retarded in particular, the same rights and responsibilities we accord all citizens and we should restrict those rights and responsibilities only when there are good reasons to do so.

IV. COMPETENCE AND THE DEFINITION OF MENTAL RETARDATION

What will count as good reasons turns on how we should understand the definition of mental retardation and its relationship to the notion of competence just explained. There are two basic problems with the definition, each of which will be taken up in turn. The first problem concerns the intrinsic aspect of the label and has its roots in the first half of the standard definition offered by the American Association on Mental Deficiency (AAMD): "Mental retardation refers to significantly subaverage general intellectual functioning . . ." ([17], p. 11). The danger here is that, without further elaboration on what 'general' means in the concept of 'general intellectual functioning', one might be tempted to think of subnormal general intellectual functioning as a *fixed state,* not open to alteration. And, when viewed as a fixed state, the label of mental retardation implies that one is broadly incompetent because one's general intellectual capacities are subaverage. Thus, one's moral status is significantly reduced in both strength and scope.

The assumption behind such a view is that one is all at once or not at all incompetent. This assumption is mistaken because it is tied to a mistaken concept of person: that they come to be all at once and remain, much like Leibniz's monads [26] and Aristotle's substances [2], hale and robust until they are destroyed and thus cease altogether to be. As H. Tristram Engelhardt has argued in the context of the abortion controversy [12] and as Stephen

Wear has argued concerning mental illness [36], such a view cannot be sustained. Instead, a more process-oriented concept of person is called for, one that recognizes that persons come to be in stages and that all of us are at any one time only more or less persons.

Put more exactly, what is called for is a task-specific concept of competence. On such a view, one could well be competent in one respect and not in another. In the case of the mentally retarded, for example, one might be competent to hold a job in a supervised setting but not in an unsupervised one. The fixed state view of mental retardation in the intrinsic sense cannot acknowledge this fundamental aspect of competence and thus of moral status. Thus, if my argument is correct, to argue that the world in which one broadly loses moral status is necessarily connected to the label of mental retardation fails because of a conceptual confusion at its very foundations. This is not a trivial point, since on the fixed state view of mental retardation there would, as we have seen already, be no reason to expend disproportionate levels of resources in educational, job-related, or housing programs for those thus afflicted since those expenditures would do no good. Moreover, many legal restrictions on the retarded are based on this view [1]. Thus, whatever opportunity one might have had to develop, with assistance, whatever intellectual capacity one possessed would be lost, leaving one worse off than one might have been and significantly disadvantaged vis-a-vis those of one's peers who were not so labeled.

The cure for the problem, it seems to me, is to reject a fixed state view of mental retardation, because one ought to reject a fixed state view of levels of intellectual capacities and the metaphysics of persons that sustains it. To do so is not to deny that some of the mentally retarded, especially those who are profoundly and severely retarded, do not have very limited horizons into which capacities might be expanded. Nonetheless, a process-oriented view provides the grounds for protecting the retarded, especially the mildly and moderately retarded (the vast bulk of the retarded population) from unfounded loss of opportunity for human development, based on a failure to recognize the concept of task-specific competency.

Happily, the more process-oriented view is reflected in the American Association on Mental Deficiency's elaboration on the classification of the retarded. In referring to the use of the terms mild, moderate, severe, and profound, the authors of the AAMD *Manual on Terminology and Classification* caution:

These terms are, of course, not absolute or static. A child classified as mildly retarded may be better served in a "trainable" class than an "educable" one; some children at the severe retardation level may function successfully in a "trainable" group; children may move up and down between categories. The level does not necessarily dictate the particular service needed, but may be helpful as one criterion in planning ([17], p. 19).

The problems surrounding the extrinsic aspects of labeling, those dealing with behavioral manifestations of mental retardation, are more difficult to root out because they are considerably more complex. These center on the second half of the AAMD's definition of mental retardation: " . . . existing concurrently with deficits in adaptive behavior . . . " ([17], p. 11). Here it is not enough, as it was above, to exchange one view for another, because of a fundamental lack of clarity concerning the causal connection between "significantly subaverage general intellectual functioning" and "deficits in adaptive behavior." The significance of this lack of clarity in the labeling of the retarded emerges when we recognize that both intellectual capacities and behavioral capacities are directly tied to the notion of task-specific competence. That is, the possession of rights and responsibilities has as its necessary conditions behavioral, as well as mental, competencies. Consider the right to vote. In most states, as I understand matters, all that is required as competency to vote are the ability to complete a voter registration form and to mark a ballot. Clearly, both mental and behavioral competencies are required to fullfill both these conditions.

What relationship between subnormal behavioral and mental deficiencies (incompetencies) is legitimated by the label of mental retardation? The answer to this question turns on clarifying the causal relationship between the two, since only those behavior deficiencies that are causally linked with mental deficiencies can be legitimately involved as reasons to limit moral status because of mental retardation. So I wish now to argue. As long as the causal relationship between the two remains murky – and what could be murkier than the phrase "existing concurrently with" – then we are deprived of reliable grounds for determining competence.

The authors of the AAMD *Manual* recognize that "Knowledge of the relationship between intelligence and adaptive behavior is still incomplete" ([17], p. 11). The problem is compounded by their explication of adaptive behavior:

Adaptive behavior is a product of the interactions of an individual's abilities and skills with the expectations of society and of the opportunities to learn ([17], p. 12).

and their insistence that both mental deficiency and behavioral deficiency are required to make an accurate diagnosis and thus label someone as mentally retarded.

There are, however, at least three separate problems that can be raised at this point. The first concerns cases of failure to diagnose mental deficiency. According to what the AAMD has to say, a child who was congenitally mentally deficient yet exhibited adaptive behavior would not be properly labeled as mentally retarded. Indeed, if only behavioral assessment were used, such a child might be labeled 'normal'. As a consequence, the child would not qualify for special education assistance to address his or her problem of mental deficiency. This consequence might well leave the child worse off, especially if his mental deficiency could be overcome or ameliorated. So a rather powerful argument could be made that the label 'normal' in such a case leads to an unjustifiable state of affairs. That one has simply adapted to one's environment with appropriate behavior does not mean that *sustained* opportunities to develop one's mental capacities are thereby secured. The solution, I think, is to have the label refer primarily to mental deficiency, so that this entitlement is not lost.

When both criteria are to be employed, then we need to attend to a second problem, the relationship between adaptive behavior and one's environment. The problem, of course, is that there are many environments and many adaptations to them. That is, there is no standard environment by which to measure "normal" behavior [21, 11]. Some have recognized this problem. Leland, for example, notes that often what counts as adaptive behavior can be quite narrowly defined.

As we observe people working with children, we find that they usually approach the child in terms of their own biases. If an aide thinks that a child is not toilet-trained, she will approach and even dress the child as though he were not. It matters little whether the child is really toilet-trained; for as long as the aide feels that he is not, and continues to approach him in this manner, the environment with which the child must cope includes a prime authority figure who feels that he is not toilet-trained ([27] p. 73).

Thus, Leland concludes, when we speak of behavior as adaptive to an environment, "it is the impression of the person in authority that becomes what the child is 'really' doing" ([27], p. 74). As a corrective, Leland has proposed that only "those specific activities essential to his survival in a particular setting" ([27], p. 74) count as adaptive behavior. To accept this solution requires that we agree on what will count as survival in a particular setting and on what activities are essential to survival. But an attempt at such agreement will come to grief on the shoals of pluralism, as Leland himself admits.

In addition to the problem of what counts conceptually as adaptive behavior, there are serious problems in applying the concept. Clausen cites a number of studies that point out that "no adequate measures of adaptive behavior are currently available" ([7], p. 52). The AAMD *Manual* admits as much and offers, as a corrective, that "measures of adaptive behavior . . . must be determined on the basis of a series of observations in many places over considerable periods of time" ([17], p. 17). Somehow, one is left to presume, this practice will ensure objectivity, but only in the sense of a consensus on what is observed. Whether the observations are accurate and whether what is observed is causally related to mental deficiency are not assured by this methodological device. Since what will count as adaptive behavior remains uncertain, the problems are compounded.

We are thus left where we started, in a muddle about the causal connection between mental deficiency and adaptive behavior. Now, this muddle is a serious one, since it leaves us unable to make the determinations of task-specific competency we require, if we are to determine on reasonable grounds the proper range of rights and responsibilities to be accorded mentally retarded individuals in virtue of having been labeled as such. Moreover, it implies that simple class membership of those labeled mentally retarded justifies loss of moral status in a broad-sweeping way. Thus, leaving this muddle unresolved opens the label of mental retardation to abuses in the sociological realm much like those Nicholas Kittrie describes in the case of those labeled mentally ill [23] and those Kindred and Rothman describe in the case of the mentally retarded [22, 35].

The way out of this muddle, I want now to suggest, is to include only those behavioral maladaptations *causally grounded* in and not simply "existing concurrently with" mental deficiency to be included in the definition of mental retardation. In this way, we can remedy the defects noted just above. In particular, ethical or policy implications regarding moral worth based solely on class membership are ruled out, because what will count as maladaptive behavior must be individualized to the person so labeled as required by this revised definition of mental retardation. Moreover, to avoid the problems Leland notes, the person to be labeled should be observed in a number of different environments by different observers. And, by counting as legitimated by the label of mental retardation only those maladaptive behaviors caused by mental deficiency, we can acquire the basis for developing more reliable, if not finally objective, measures. In short, what the label of mental retardation legitimates is an individuated assessment of mental and behavioral deficiencies, similar to what Lawton and Nahemov have termed the ecological model of competence [25]. Ronald Conley sums up this view nicely:

Social incompetence may result from mental subnormality, but cannot be considered as diagnostic of this condition, unless a *one-to-one relationship* between the degree of social competence and the degree of mental subnormality can be demonstrated ([8], p. 9).

In short, the label of mental retardation ought to focus primarily on its intrinsic dimensions and only on those extrinsic aspects causally rooted in those intrinsic dimensions. What one avoids by doing so are arbitrary and unjustifiable losses of moral status, i.e., those based on simple class membership, while still gaining for the mentally retarded the special assistance they require and to which, as argued above, they are entitled. Indeed, as Conley points out, a narrower definition, one based on isomorphic mapping of intellectual functioning onto behavioral deficits "would be more definitive of the population that is likely to need assistance" ([8], p. 9).

The normative impact of this more restrictive understanding of the label of mental retardation is the following. It is first in its intrinsic sense and then in this refined extrinsic sense that the label necessarily implies that a world is lost, namely the world of "normal" intellectual functioning. But, on this view moral status is not lost across the board, given the arguments above against the fixed state view of mental retardation and those in favor of task-specific competency. Thus, while moral status is necessarily lost, the nature and extent of that loss are restricted to a determination made on an individual basis, by developing an individuated map of mental and behavioral incompetencies and tying them to specific rights and responsibilities. An "across-the-board" loss of moral status, of rights and responsibilities, is thus not legitimated by the label of mental retardation, given the reformulation of the definition of mental retardation set out above. And labeling the retarded, as I understand it here, would not legitimate an argument from social incompetence to mental incompetence to broadly diminished moral status. Instead, only arguments from mental incompetence alone or from mental incompetence tied one-to-one to behavioral maladaptation to diminished moral status are legitimated by the label. Arguments from social incompetence alone to diminished moral status are ruled out as well, if one appeals to the label of mental retardation as legitimation for them. If instead one argues from social incompetence as itself a label – applying obviously to many more than just retarded individuals – then diminished moral status can perhaps be implied. Such a move would, however, have nothing to do by way of legitimation with the label of mental retardation. Whether it can be justified at all is a question beyond the scope of this paper.

V. LABELING AND THE "CLINICAL PERSPECTIVE"

The position regarding labeling I have advocated above comes out as a mix of what Jane Mercer has called the 'clinical perspective', and the 'social system perspective' [31]. In closing I want to consider some of the advantages of my more mixed understanding of labeling over her social-system-oriented approach.

According to the social system model mental retardation is an "achieved status in the social system" and is specific to that system. That is, one is normal or retarded according to whether one "meets the expectations of the social system in which he is operating" ([31], p. 36). In contrast, the clinical model views mental retardation as a pathological condition, cross-cultural in nature, that exists whether or not it is diagnosed ([31], p. 36). Now, it strikes me that the social system perspective has the same problems we have found to be associated with the notion of adaptive behavior. For example, if one's behavior satisfies the expectations of one's society yet one is mentally deficient, then according to this perspective, it would not be accurate to label such an individual as mentally retarded; in consequence, one would lose one's ticket for special treatment and in all likelihood wind up worse off vis-a-vis development of intellectual function. On the account with which we began, such a state of affairs would be unjustified. Moreover, what will count as "normal" behavior will be arbitrary because of factors like those identified by Leland. Thus, many will suffer arbitrary losses of moral status by being inaccurately labeled as retarded.

There are, then, serious shortcomings in what at first appears an attractive model. Indeed, these shortcomings are, in my view, serious enough in their ethical implications to warrant rejection of the social system perspective as alone adequate, since it falls prey to the criticism above concerning the concept of maladaptive behavior. Instead, it should be wed to the clinical perspective, rather than separated from it. By doing so, we bring to the social perspective what I regard as a very powerful aspect of the clinical perspective, whose value-laden character [13] now emerges in its insistence that individuals be assessed as such and not simply in terms of how they fit or fail to fit "expectations of the social system in which [one] is operating" ([31], p. 36). In short, the social perspective itself needs to be individuated, if it is to avoid the problems already found to characterize both the intrinsic and extrinsic aspects of the label of mental retardation and their proper relation.

VI. SUMMARY

In this paper I have examined some of the fundamental dimensions of labeling the mentally retarded. I began with considerations of how we should understand that labeling rightly gains a world in which special treatment of the retarded is a matter of obligation. I then asked whether that obligation of special treatment must be always be purchased at the price of lost moral status. To answer this, the main question of this paper, I undertook an analysis of the notion of a handicap, in the process of which intrinsic and extrinsic aspects of the label were distinguished. In the case of mental retardation, the former refers to mental deficiencies and the latter primarily to their behavioral manifestations. I then argued that a fixed state view of retardation should be abandoned in favor of a process view, in order to avoid some ethical problems associated with the intrinsic aspect of the label. The most serious problems associated with the extrinsic aspect of labeling, our attitudes toward those labeled as retarded and the social practices those attitudes create and sustain, can be avoided, I argue, by resolving the conceptual confusion about the causal relationship between mental and social incompetence by individuating the connection. Thus, the label of mental retardation legitimates loss of moral status not on the basis of class membership but only on an individual basis, based either in the intrinsic aspect of the label or in a one-to-one mapping of the extrinsic onto the intrinsic aspects of the label. Thus, in Mercer's terms, a mixed clinical-perspective/social-system view of labeling has been defended here.

The upshot of all of this for how we should understand the ethical dimensions of labeling the retarded is that the label retains its character as a ticket for special treatment and the grounds for this aspect of the label are clarified: they concern mental deficiencies and the behavior deficiencies they cause. Thus, the label no longer necessarily implies in a broad and far-reaching manner the loss of moral status and thus a condition of greatly diminished moral worth. Such an implication is based on fundamental conceptual confusion about mental retardation. Thus, the grounds for any necessary connection between the label and loss of moral status are quite narrow; they are, indeed, the same as those that legitimate special treatment. And, in light of the concepts of task-specific competence and task-specific incompetence whose character I have outlined above, any loss of moral status must be determined on an individual, case-by-case basis. Simple class membership, therefore, implies nothing for loss of moral status when one is (accurately) labeled mentally retarded. Finally, though this surely constitutes the loss of a world, that loss is not as wide-ranging and as inevitably arbitrary as the loss to which the mentally retarded have been historically subjected in our society, i.e., on the basis of simple class membership. Thus, there is no good reason that the world gained for the mentally retarded by being so labeled is to be purchased at the devastating price that many of the retarded have been forced to pay because of our failure to understand in a reliable manner the conceptual and ethical dimensions of this crucially important diagnostic and social label.

Georgetown University School of Medicine
Washington, D. C.

NOTES

[1] What I am offering here is not entirely faithful to the Rawlsian text. Indeed, it might be taken as corrective of it. That is, I am attempting a further refinement of the concept of the rational contractor in the original position, by suggesting that those in an initial position of equality would have a fundamental interest in maintaining that equality, compatible with a mutually maximal equality of others. Hence the emphasis on "sustained opportunity" for human development. This, I take it, is the force of Rawls' characterization of those in the original position as rational and mutually disinterested.

This formulation also gives content to the rather opaque concept of 'expectations' in Rawls' treatment of the Difference Principle. That is, the notion of sustained opportunity for human development goes beyond what appear to be psychological claims to conceptual claims about the necessary conditions for creating and maintaining a just social order – precisely the concern of the rational contractors in the original position. To satisfy this demand on a theory of justice, the language of expectations will not do. We are, instead, forced back to a philosophical conception of human nature in the moral and political order. Thus, in a Kantian mode, one might say – the argument remains to be explored in detail – that a necessary condition of a *just* community, and not simply a peaceable one, is that there be sustained opportunities for human development and not simply access to them (all that the peaceable community need address) [14, 32].

[2] It may well be that there are no reliable measures of the competencies required to be an adequate parent. If that is the case, then even this narrowly based view of sterilization would have to be abandoned.

BIBLIOGRAPHY

[1] Areen, J.: 1978, *Cases and Materials on Family Law,* The Foundation Press, Mineola, New York, pp. 251–261.

[2] Aristotle, 1941, *Metaphysics*: in R. McKeon (ed.), *The Basic Works of Aristotle,* Random House, New York, pp. 689–926.

[3] Austin, J. L.: 1961, *Philosophical Papers,* J. O. Urmson and G. J. Warnock (eds.), Oxford University Press, New York. See especially Chapter 10, 'Performative Utterance', pp. 223–252.

[4] Austin, J. L.: 1968, *How To Do Things With Words,* J. O. Urmson (ed.), Oxford University Press, New York.

[5] Blackstone, W. T.: 1968, 'Equality and Human Rights', *The Monist* **52**, 627–628.

[6] Callahan, D.: 1976, 'Biomedical Progress and the Limits of Human Health', in R. M. Veatch and R. Branson (eds.), *Ethics and Health Policy,* Ballinger Publishing Company, Cambridge, Massachusetts, pp. 157–165.

[7] Clausen, J.: 1972, 'Quo Vadis, AAMD?', *Journal of Special Education* **6**, 51–60.

[8] Conley, R. W.: 1973, *The Economics of Mental Retardation,* Johns Hopkins University Press, Baltimore, Maryland.

[9] Cooke, R. E.: 1978, 'Mentally Handicapped.' in W. T. Reich (ed.), *Encyclopedia of Bioethics,* Macmillan Publishing Co., The Free Press, New York, pp. 1108–1114.

[10] Daniels, N.: 1979, 'Rights to Health Care and Distributive Justice: Programmatic Worries', *Journal of Medicine and Philosophy,* **4**, 174–191.

[11] Edgerton, R. B.: 1970, 'Mental Retardation in Non-Western Societies: Toward a Cross-Cultural Perspective on Incompetence', in H. Carl Haywood (ed.), *Social-Cultural Aspects of Mental Retardation,* Appleton-Century-Crofits, New York, pp. 523–559.

[12] Engelhardt, H. T., Jr.: 1974, 'The Ontology of Abortion', *Ethics* **84**, 217–234.

[13] Engelhardt, H. T., Jr. and Spicker, S. F. (eds.): 1975, *Evaluation and Explanation in the Biomedical Sciences,* D. Reidel Publishing Co., Dordrecht, Holland.

[14] Engelhardt, H. T., Jr.: 1981, 'Health Care Allocations: Responses to the Unjust, the Unfortunate, and the Undesirable', in E. Shelp (ed.), *Justice and Health Care,* D. Reidel Publishing Co., Dordrecht, Holland, pp. 121–137.

[15] Fried, C.: 1976, 'Equality and Rights in Health Care', *Hastings Center Report* **6**, 29–34.

[16] Green, R.: 1976, 'Health Care and Justice in Contract Theory Perspective', in R. M. Veatch and R. Branson (eds.), *Ethics and Health Policy,* Ballinger Publishing Company, Cambridge, Massachusetts, pp. 111–126.

[17] Grossman, H. J. (ed.): 1977, *Manual on Terminology and Classification in Mental Retardation,* American Association on Mental Retardation, Washington, D. C.

[18] Guskin, S. L.: 1978, 'Theoretical and Empirical Strategies for the Study of Labeling of Mentally Retarded Persons', in N. R. Ellis (ed.), *International Review of Research in Mental Retardation,* Vol. 9, Academic Press, New York, pp. 127–158.

[19] Hapgood, J. S. *et al.:* 1979, 'Sterilization of the Mentally Handicapped', *Lancet* **11**, 685–686.

[20] Hauerwas, S.: 1977, *Truthfulness and Tragedy,* Univ. of Notre Dame Press, Notre Dame, Indiana.

[21] Haywood, H. D. (ed.): 1970, *Social-Cultural Aspects of Mental Retardation,* Appleton-Century-Crofts, New York.

[22] Kindred, M.: 1982, 'The Legal Rights of Mentally Retarded Persons in Twentieth Century America', in this volume, pp. 185–208.

[23] Kittrie, N.: 1978, 'Labeling in Mental Illness: Legal Aspects', in W. T. Reich

(ed.), *Encyclopedia of Bioethics,* Macmillan Publishing Co., The Free Press, New York, pp. 1102–1108.

[24] Largey, G.: 1977, 'Reversible Sterilization', *Society* 14, 57–59.

[25] Lawton, M. D. and Nahemov, L.: 1973, 'Ecology and the Aging Process', in C. Eisdorfer and M. P. Lawton (eds.), *Psychology of Adult Development and Aging,* American Psychological Association, Washington, D. C.

[26] Leibniz, G. W.: 1976, 'The Monadology', in L. E. Loemker (ed.), *Leibniz's Philosophical Papers and Letters,* D. Reidel Publishing Co., Dordrecht, Holland, pp. 643–653.

[27] Leland, H.: 1972, 'Mental Retardation and Adaptive Behavior', *Journal of Special Education* 6, 71–80.

[28] Ludlow, J. R. and Allen, A. M.: 1979, 'The Effect of Early Intervention and Pre-School Stimulus on the Development of the Down's Syndrome Child', *Journal of Mental Deficiency Research* 23, 29–44.

[29] Macmillan, D. L., Jones, R. L., and Aloia, G. F.: 1947, 'The Mentally Retarded Label: A Theoretical Analysis and Review of Research', *American Journal of Mental Deficiency* 79, 241–261.

[30] Mason, B. G., Manolascino, F. J., and Galvin, L.: 1976, 'The Right to Treatment for Mentally Retarded Persons: An Evolving Legal and Scientific Interface', *Creighton Law Review* 10, 124–169.

[31] Mercer, J. R.: 1973, *Labeling the Mentally Retarded,* Univ. of California Press, Berkeley.

[32] Nozick, R.: 1974, *Anarchy, State, and Utopia,* Basic Books, New York.

[33] Potter, R. B.: 1977, 'Labeling the Mentally Retarded: The Just Allocation of Theory', in S. Reiser *et al.* (eds.), *Ethics in Medicine: Historical Perspectives and Contemporary Concerns,* MIT Press, Cambridge, Mass., pp. 626–630.

[34] Rawls, J.: 1971, *A Theory of Justice,* Harvard University Press, Cambridge, Mass.

[35] Rothman, D.: 1982, 'Who Speaks for the Retarded?', in this volume, pp. 223–233.

[36] Wear, S.: 1980, 'Mental Illness and Moral Status', *Journal of Medicine and Philosophy* 5, 292–312.

ROBERT L. HOLMES

LABELING THE MENTALLY RETARDED: A REPLY TO LAURENCE B. McCULLOUGH

Philosophers understandably have a fondness for conceptual muddles; once uncovered and set straight, they cause seemingly intractable problems to vanish, ill-conceived theories to collapse and light to shine where darkness had reigned – all without one's having to leave one's philosophical armchair.

So it is with particular interest that we examine Laurence McCullough's claim that the root problem in the sweeping loss of rights and responsibilities supposedly legitimated by the label 'mentally retarded' "is a conceptual muddle surrounding the definition of mental retardation" ([1], p. 99). His underlying concern is with whether one can accord special treatment to the handicapped without at the same time causing them to lose moral status, in the sense of full representation of rights and responsibilities. Must the world "gained" always be accompanied by a world "lost"?

I

Let us begin by asking what precisely the alleged muddle is. To do this we need to take account of what McCullough calls "the intrinsic and extrinsic value dimensions of labels for handicapping conditions," but which we may, for short, refer to simply as intrinsic and extrinsic handicaps. An intrinsic handicap is one that one simply *has*, by chance or nature; an extrinsic handicap is one that has been *imposed* upon one by someone else (McCullough adds that it has been imposed consciously or deliberately, but that seems an unnecessary qualification).

Now McCullough claims that "when we speak of fellow human beings as being handicapped in the first sense . . . we in very subtle ways also impose a handicap in the second sense. That is, our attitudes, value judgments and practices often place the handicapped at additional disadvantage" ([1], 103). Why does he think this? Three possibilities suggest themselves. He may believe (1) that 'x is handicapped' *means*, at least in part, or is taken to mean, something like 'x is of lower moral status'; or (2) that people *associate* diminished moral status with being handicapped, even though the word 'handicapped' (which actually occupies most of McCullough's attention, rather than the label 'mentally retarded') may not have that as part of its

L. Kopelman and J. C. Moskop (eds.), Ethics and Mental Retardation, 119–123.

conceptual meaning; or (3) that people simply *respond* to the handicapped (or, if you like, to the word 'handicapped') in ways which disadvantage them, whether or not they believe that being handicapped signifies a diminished moral status. Any of these could plausibly be thought to impose, or to lead to the imposition of, an external handicap.

McCullough seems to lean toward (1) when he says that "... our ordinary use of the language of being handicapped embraces complex value judgments with important implications for how we treat those with handicaps" ([1], 104). If the handicapped are, by definition, of lower moral status, this would be eminently true. On the other hand, he says such things as that "Reasonable or justifiable attitudes toward the handicapped will avoid ethical judgments that arbitrarily restrict the handicapped person's ability to develop his or her potential" ([1], 105), which would suggest (2) or (3).

Whether any of these represent an illegitimate imposition of external handicaps depends, of course, upon whether the meaning, attitudes and responses they respectively signify are warranted. Consider (3), for example. It is true that people often do respond to the handicapped in ways which unjustly disadvantage them, and no one would disagree that this is wrong. But this is just another way of saying that the handicapped are often treated badly. It does not in any way indicate a conceptual confusion accounting for such treatment. People can understand full well the nature of handicaps and still behave deplorably towards those who have them. So there is no apparent conceptual confusion in (3).

What of (2)? Is it a mistake to associate handicaps with diminished moral status? The tenor of McCullough's discussion suggests that he thinks it is, but the analysis he gives implies otherwise. For he subscribes to what he calls an Ethical Principle of Diminished Competence. or what we may refer to as EPDC for short. It says that "as one's capacities requisite for the exercise of rights and responsibilities diminish, so too do one's rights and responsibilities" ([1], 107). Given McCullough's further claim that these capacities are necessary conditions of moral status, it follows that one's moral status is undermined to the extent that these capacities are diminished. Since, by definition, to be handicapped is to have significantly diminished capacities of one sort or another, it follows that to be handicapped is to have a diminished moral status. If that is so, there is no confusion in associating diminished moral status with being handicapped, since in fact the two are connected. McCullough comes close to acknowledging this when he says, after stressing that assessment of diminished capacity should be made on a case by case basis rather than by membership in the class of the handicapped, that "Thus, *while moral status is necessarily lost,* the nature and extent of that loss are restricted to a determination made on an individual basis . . . " [italics added] ([1], 113). This is not to say that the claim made by (2) is correct; only that it cannot provide the basis for the alleged conceptual confusion that McCullough has in mind, since he himself is committed to it.

This leaves (1) as the possible source of confusion. And there is one mistake which McCullough says may accompany use of the notion of intrinsic handicap. It is that of supposing that a handicap is a fixed state, incapable of amelioration by effort or training. McCullough spends considerable time discussing this view, but he does not show any necessary connection between it and the idea of an intrinsic handicap; in fact, the only actual definition he considers – the definition of mental retardation by the American Association on Mental Deficiency – expressly repudiates the view. Though he contends that the fixed state view is "tied to a mistaken concept of a person" ([1], 108), the mistake represented by such a view would seem to be empirical rather than conceptual. Whether retarded or not, people either can or cannot improve their capabilities, and the determination of whether they can do so can be made only by the appropriate observations or tests. If someone argued that experience is irrelevant to this question, and concluded a priori from the concept of a person that the handicapped cannot improve their capabilities, *that*, of course, would be a mistake, and would have some claim to be called a conceptual confusion. But save for some unexplained references to Aristotle and Leibniz, McCullough does not cite anyone who has ever held this view, and does not attempt to detail the supposed connection between the fixed-state view and the concept of a person.

Apart from cautioning against building the fixed state view into the meaning of 'handicapped', it is difficult to see what McCullough thinks is a conceptual confusion, making ascriptions of handicaps necessarily pejorative. There *could*, of course, be definitions which had this consequence; one could define 'x is handicapped' to mean 'x is inferior' or 'x is sub-human'. But McCullough does not consider any such definitions nor does he make a case for supposing that people generally take the word 'handicapped' to mean such things. The one definition he does consider, of mental retardation, does not do this. If it did, the result would be a bad definition, one which reflected indefensible moral and factual beliefs. But it would not even then necessarily represent a *conceptual* confusion.

There remains, however, one further possiblity. It may be that the main mistake McCullough has in mind is the generalization from mere membership in the class of handicapped persons to the conclusion that there is a "loss of moral status in a broad-sweeping way" ([1], 112). This apparently means that some people conclude that just because an individual is handicapped, whatever the handicap and however minor it may be, he or she is then deemed to have lost a vast range of rights and responsibilities. This, of course, is a mistake, and to the extent people do make it McCullough is correct to point that fact out. And his proposed alternative, that determinations of loss of moral status should be made on a case by case basis, similarly makes sense. But he gives us no reason to believe that such a mistake, if indeed it is made, has anything to do with a conceptual confusion in the definition of mental retardation. The likelier explanation is thoughtlessness, ignorance, insensitivity and in some cases indefensible moral veiws.

If the preceding is correct, nothing that warrants being called a conceptual confusion emerges from his discussion, and his explanation of why imputing intrinsic handicaps involves imposing external handicaps reduces pretty much to saying that people often mistreat the handicapped; there is little more in the claim that it is wrong to impute a loss of moral status 'in a broad sweeping way' solely on the basis of class membership in the handicapped. Finally, though he does not seem to intend this, McCullough's own analysis commits him to saying that there *is* a loss of moral status bound up in according special treatment to the handicapped. The EPDC, conjoined with the other views McCullough apparently holds, entails it. The most that he can say is that we should not assess that loss "in a broad sweeping way."

II

Finally, I want to return briefly to the EPDC. I have said that if one subscribes to that principle, it supports the position that it is correct to associate diminished moral status – that is, loss of rights and responsibilities – with being handicapped. I now want to suggest that this principle should be rejected.

As far as responsibilities are concerned, the loss of capabilities entailed by one's being handicapped need result in no diminution of responsibilities. It most often will, and should, result in a *change* in responsibilities. It is precisely because the handicapped have often been thought incapable of assuming significant responsibilities that they have been prevented from realizing their potential. The need is to give them responsibilities they can handle and to relieve them of those they cannot – in short, to give them suitable responsibilities rather than fewer responsibilities.

With regard to rights, there is at least one reason to reject the view of

EPDC that the loss or absence of capacities to exercise a right signifies a loss of that right. This may often be true of legal rights, but it is not necessarily true of moral rights, particularly so-called human rights. Rights like the right not to be harmed do not need to be "exercised"; one simply has them, irrespective of what one does (excluding cases in which one's bad conduct is sometimes thought to involve a forfeiture of the right). They do not presuppose any particular capacities and are not lost in the absence of those capacities. Even where rights do need to be exercised to be fully enjoyed – like rights to free speech and assembly, which may be moral as well as legal – there is no reason to say that they are lost if one cannot exercise them. There might be such reason if the fixed state view of handicaps, which McCullough rejects, were accepted. For then the deficiencies in question would be incapable in principle of improvement, hence the right incapable of being exercised. But even where rights are incapable of *ever* being exercised – I will never play professional football if I have lost a leg, or be a concert pianist if I have lost both hands – to deny the existence of the right opens the way to denying moral status in a far more serious way than McCullough discusses; it opens the way to regarding not only the handicapped, but also the infirm and the elderly – and *all* of us suffer diminished capacities as we age – as inferior or sub-human persons. If a class of persons are thought not to have rights, it is only one further step, as the experience of Nazi Germany attests, to regarding them as disposable.

I submit, therefore, that we can say everything we want to say about the nature of handicaps, and about the moral principles governing the proper treatment of the handicapped, without denying that the handicapped possess the same basic human rights as everyone else.

University of Rochester
Rochester, New York

BIBLIOGRAPHY

[1] McCullough, L. B.: 1983, *The World Gained and The World Lost: Labeling the Mentally Retarded*, in this volume, pp. 99–118.

SECTION III

THEOLOGY AND PHILOSOPHY OF RELIGION

ROBERT M. ADAMS

MUST GOD CREATE THE BEST?

I

Many philosophers and theologians have accepted the following proposition:

(*P*) If a perfectly good moral agent created any world at all, it would have to be the very best world that he could create.

The best world that an omnipotent God could create is the best of all logically possible worlds. Accordingly, it has been supposed that if the actual world was created by an omnipotent, perfectly good God, it must be the best of all logically possible worlds.

In this paper I shall argue that ethical views typical of the Judeo-Christian religious tradition do not require the Judeo-Christian theist to accept (*P*). He must hold that the actual world is a good world. But he need not maintain that it is the best of all possible worlds, or the best world that God could have made.[1]

The position which I am claiming that he can consistently hold is that *even if* there is a best among possible worlds, God could create another instead of it, and still be perfectly good. I do not in fact see any good reason to believe that there is a best among possible worlds. Why can't it be that for every possible world there is another that is better? And if there is no maximum degree of perfection among possible worlds, it would be unreasonable to blame God, or think less highly of His goodness, because He created a world less excellent than He could have created.[2] But I do not claim to be able to prove that there is no best among possible worlds, and in this essay I shall assume for the sake of argument that there is one.

Whether we accept proposition (*P*) will depend on what we believe are the requirements for perfect goodness. If we apply an act-utilitarian standard of moral goodness, we will have to accept (*P*). For by act-utilitarian standards it is a moral obligation to bring about the best state of affairs that one can. It is interesting to note that the ethics of Leibniz, the best-known advocate

Originally published in *The Philosophical Review* 81 (1972), 317–332. Reprinted by permission.

L. Kopelman and J. C. Moskop (eds.), Ethics and Mental Retardation, 127–140.

of (*P*), is basically utilitarian ([1], pp. 210–218). In his *Theodicy* (Part I, Section 25) he maintains, in effect, that men, because of their ignorance of many of the consequences of their actions, ought to follow a rule-utilitarian code, but that God, being omniscient, must be a perfect act utilitarian in order to be perfectly good.

I believe that utilitarian views are not typical of the Judeo-Christian ethical tradition, although Leibniz is by no means the only Christian utilitarian. In this essay I shall assume that we are working with standards of moral goodness which are not utilitarian. But I shall not try either to show that utilitarianism is wrong or to justify the standards that I take to be more typical of Judeo-Christian religious ethics. To attempt either of these tasks would unmanageably enlarge the scope of the paper. What I can hope to establish here is therefore limited to the claim that the rejection of (*P*) is consistent with Judeo-Christian religious ethics.

Assuming that we are not using utilitarian standards of moral goodness, I see only two types of reason that could be given for (*P*). (1) It might be claimed that a creator would necessarily wrong someone (violate someone's rights), or be less kind to someone than a perfectly good moral agent must be, if he knowingly created a less excellent world instead of the best that he could. Or (2) it might be claimed that even if no one would be wronged or treated unkindly by the creation of an inferior world, the creator's choice of an inferior world must manifest a defect of character. I will argue against the first of these claims in Section II. Then I will suggest, in Section III, that God's choice of a less excellent world could be accounted for in terms of His grace, which is considered a virtue rather than a defect of character in Judeo-Christian ethics. A counterexample, which is the basis for the most persuasive objections to my position that I have encountered, will be considered in Sections IV and V.

II

Is there someone *to* whom a creator would have an obligation to create the best world he could? Is there someone whose rights would be violated, or who would be treated unkindly, if the creator created a less excellent world? Let us suppose that our creator is God, and that there does not exist any being, other than Himself, which He has not created. It follows that if God has wronged anyone, or been unkind to anyone, in creating whatever world He has created, this must be one of His own creatures. To which of His creatures, then, might God have an obligation to create the best of all possible worlds? (For that is the best world He could create.)

Might He have an obligation to the creatures in the best possible world, to create them? Have they been wronged, or even treated unkindly, if God has created a less excellent world, in which they do not exist, instead of creating them? I think not. The difference between actual beings and merely possible beings is of fundamental moral importance here. The moral community consists of actual beings. It is they who have actual rights, and it is to them that there are actual obligations. A merely possible being cannot be (actually) wronged or treated unkindly. A being who never exists is not wronged by not being created, and there is no obligation to any possible being to bring it into existence.

Perhaps it will be objected that we believe we have obligations to future generations, who are not yet actual and may never be actual. We do say such things, but I think what we mean is something like the following. There is not merely a logical possibility, but a probability greater than zero, that future generations will really exist; and *if* they will in fact exist, we will have wronged them if we act or fail to act in certain ways. On this analysis we cannot have an obligation to future generations to bring them into existence.

I argue, then, that God does not have an obligation to the creatures in the best of all possible worlds to create them. If God has chosen to create a world less excellent than the best possible, He has not thereby wronged any creatures whom He has chosen not to create. He has not even been unkind to them. If any creatures are wronged, or treated unkindly, by such a choice of the creator, they can only be creatures that exist in the world He has created.

I think it is fairly plausible to suppose that God could create a world which would have the following characteristics:

(1) None of the individual creatures in it would exist in the best of all possible worlds.

(2) None of the creatures in it has a life which is so miserable on the whole that it would be better for that creature if it had never existed.

(3) Every individual creature in the world is at least as happy on the whole as it would have been in any other possible world in which it could have existed.

It seems obvious that if God creates such a world He does not thereby wrong any of the creatures in it, and does not thereby treat any of them with less than perfect kindness. For none of them would have been benefited by His creating any other world instead.[3]

If there are doubts about the possibility of God's creating such a world, they will probably have to do with the third characteristic. It may be worth while to consider two questions, on the supposition (which I am not endorsing) that no possible world less excellent than the best would have characteristic (3), and that God has created a world which has characteristics (1) and (2) but not (3). In such a case must God have wronged one of His creatures? Must He have been less than perfectly kind to one of His creatures?

I do not think it can reasonably be argued that in such a case God must have wronged one of His creatures. Suppose a creature in such a case were to complain that God had violated its rights by creating it in a world in which it was less happy on the whole than it would have been in some other world in which God could have created it. The complaint might express a claim to special treatment: "God ought to have created *me* in more favorable circumstances (even though that would involve His creating some *other* creature in less favorable circumstances than He could have created it in)." Such a complaint would not be reasonable, and would not establish that there had been any violation of the complaining creature's rights.

Alternatively, the creature might make the more principled complaint, "God has wronged me by not following the principle of refraining from creating any world in which there is a creature that would have been happier in another world He could have made." This also is an unreasonable complaint. For if God followed the stated principle, He would not create any world that lacked characteristic (3). And we are assuming that no world less excellent than the best possible would have characteristic (3). It follows that if God acted on the stated principle He would not create any world less excellent than the best possible. But the complaining creature would not exist in the best of all possible worlds; for we are assuming that this creature exists in a world which has characteristic (1). The complaining creature, therefore, would never have existed if God had followed the principle that is urged in the complaint. There could not possibly be any advantage to this creature from God's having followed that principle; and the creature has not been wronged by God's not following the principle. (It would not be better for the creature if it had never existed; for we are assuming that the world God created has characteristic (2).)

The question of whether in the assumed case God must have been unkind to one of His creatures is more complicated than the question of whether He must have wronged one of them. In fact it is too complicated to be discussed adequately here. I will just make three observations about it. The first is that it is no clearer that the best of all possible worlds would possess characteristic (3) than that some less excellent world would possess it. In fact it has often been supposed that the best possible world might not possess it. The problem we are now discussing can therefore arise also for those who believe that God has created the best of all possible worlds.

My second observation is that if kindness to a person is the same as a tendency to promote his happiness, God has been less than perfectly (completely, unqualifiedly) kind to any creature whom He could have made somewhat happier than He has made it. (I shall not discuss here whether kindness to a person is indeed the same as a tendency to promote his happiness; they are at least closely related.)

But in the third place I would observe that such qualified kindness (if that is what it is) toward some creatures is consistent with God's being perfectly good, and with His being very kind to all His creatures. It is consistent with His being very kind to all His creatures because He may have prepared for all of them a very satisfying existence even though some of them might have been slightly happier in some other possible world. It is consistent with His being perfectly good because even a perfectly good moral agent may be led, by other considerations of sufficient weight, to qualify his kindness or beneficence toward some person. It has sometimes been held that a perfectly good God might cause or permit a person to have less happiness than he might otherwise have had, in order to punish him, or to avoid interfering with the freedom of another person, or in order to create the best of all possible worlds. I would suggest that the desire to create and love all of a certain group of possible creatures (assuming that all of them would have satisfying lives on the whole) might be an adequate ground for a perfectly good God to create them, even if His creating *all* of them must have the result that some of them are less happy than they might otherwise have been. And they need not be the best of all possible creatures, or included in the best of all possible worlds, in order for this qualification of His kindness to be consistent with His perfect goodness. The desire to create *those* creatures is as legitimate a ground for Him to qualify His kindness toward some, as the desire to create the best of all possible worlds. This suggestion seems to me to be in keeping with the aspect of the Judeo-Christian moral ideal which will be discussed in Section III.

These matters would doubtless have to be discussed more fully if we were considering whether the *actual* world can have been created by a perfectly good God. For our present purposes, however, enough may have been said – especially since, as I have noted, it seems a plausible assumption that God could make a world having characteristics (1), (2), and (3). In that case He could certainly make a less excellent world than the best of all possible worlds without wronging any of His creatures or failing in kindness to any of them. (I have, of course, *not* been arguing that there is *no* way in which God could wrong anyone or be less kind to anyone than a perfectly good moral agent must be.)

III

Plato is one of those who held that a perfectly good creator would make the very best world he could. He thought that if the creator chose to make a world less good than he could have made, that could be understood only in terms of some defect in the creator's character. Envy is the defect that Plato suggests ([3], 29E–30A). It may be thought that the creation of a world inferior to the best that he could make would manifest a defect in the creator's character even if no one were thereby wronged or treated unkindly. For the perfectly good moral agent must not only be kind and refrain from violating the rights of others, but must also have other virtues. For instance, he must be noble, generous, high-minded, and free from envy. He must satisfy the moral ideal.

There are differences of opinion, however, about what is to be included in the moral ideal. One important element in the Judeo-Christian moral ideal is *grace*. For present purposes, grace may be defined as a disposition to love which is not dependent on the merit of the person loved. The gracious person loves without worrying about whether the person he loves is worthy of his love. Or perhaps it would be better to say that the gracious person sees what is valuable in the person he loves, and does not worry about whether it is more or less valuable than what could be found in someone else he might have loved. In the Judeo-Christian tradition it is typically believed that grace is a virtue which God does have and men ought to have.

A God who is gracious with respect to creating might well choose to create and love less excellent creatures than He could have chosen. This is not to suggest that grace in creation consists in a preference for imperfection as such. God could have chosen to create the best of all possible creatures, and still have been gracious in choosing them. God's graciousness in creation does not imply that the creatures He has chosen to create must be less excellent than the best possible. It implies, rather, that even if they are the best possible creatures, that is not the ground for His choosing them. And it implies that there is nothing in God's nature or character which would require Him to act on the principle of choosing the best possible creatures to be the object of His creative powers.

Grace, as I have described it, is not part of everyone's moral ideal. For instance, it was not part of Plato's moral ideal. The thought that it may be the expression of a virtue, rather than a defect of character, in a creator, *not* to act on the principle of creating the best creatures he possibly could, is quite foreign to Plato's ethical viewpoint. But I believe that thought is not at all foreign to a Judeo-Christian ethical viewpoint.

This interpretation of the Judeo-Christian tradition is confirmed by the religious and devotional attitudes toward God's creation which prevail in the tradition. The man who worships God does not normally praise Him for His moral rectitude and good judgment in creating *us*. He thanks God for his existence as for an undeserved personal favor. Religious writings frequently deprecate the intrinsic worth of human beings, considered apart from God's love for them, and express surprise that God should concern Himself with them at all.

> When I look at thy heavens, the work of thy fingers, the moon and the stars which thou hast established;
> What is man that thou art mindful of him, and the son of man that thou dost care for him?
> Yet thou hast made him little less than God, and dost crown him with glory and honor.
> Thou hast given him dominion over the works of thy hands; thou hast put all things under his feet [Psalm 8: 3–6].

Such utterances seem quite incongruous with the idea that God created us because if He had not He would have failed to bring about the best possible state of affairs. They suggest that God has created human beings and made them dominant on this planet although He could have created intrinsically better states of affairs instead.

I believe that in the Judeo-Christian tradition the typical religious attitude (or at any rate the attitude typically encouraged) toward the fact of our existence is something like the following. "I am glad that I exist, and I thank God for the life He has given me. I am also glad that other people exist, and I thank God for them. Doubtless there could be more excellent creatures than we. But I believe that God, in His grace, created us and loves us; and I accept that gladly and gratefully." (Such an attitude need not be complacent;

for the task of struggling against certain evils may be seen as precisely a part of the life that the religious person is to accept and be glad in.) When people who have or endorse such an attitude say that God is perfectly good, we will not take them as committing themselves to the view that God is the kind of being who would not create any other world than the best possible. For they regard grace as an important part of perfect goodness.

IV

On more than one occasion when I have argued for the positions I have taken in Sections II and III above, a counterexample of the following sort has been proposed. It is the case of a person who, knowing that he intends to conceive a child and that a certain drug invariably causes severe mental retardation in children conceived by those who have taken it, takes the drug and conceives a severely retarded child. We all, I imagine, have a strong inclination to say that such a person has done something wrong. It is objected to me that our moral intuitions in this case (presumably including the moral intuitions of religious Jews and Christians) are inconsistent with the views I have advanced above. It is claimed that consistency requires me to abandon those views unless I am prepared to make moral judgments that none of us are in fact willing to make.

I will try to meet these objections. I will begin by stating the case in some detail, in the most relevant form I can think of. Then I will discuss objections based on it. In this section I will discuss an objection against what I have said in Section II, and a more general objection against the rejection of proposition (*P*) will be discussed in Section V.

Let us call this Case (*A*). A certain couple become so interested in retarded children that they develop a strong desire to have a retarded child of their own – to love it, to help it realize its potentialities (such as they are) to the full, to see that it is as happy as it can be. (For some reason it is impossible for them to *adopt* such a child.) They act on their desire. They take a drug which is known to cause damaged genes and abnormal chromosome structure in reproductive cells, resulting in severe mental retardation of children conceived and born. They lavish affection on the child. They have ample means, so that they are able to provide for special needs, and to insure that others will never be called on to pay for the child's support. They give themselves unstintedly, and do develop the child's capacities as much as possible. The child is, on the whole, happy, though incapable of many of the higher intellectual, aesthetic, and social joys. It suffers some pains and frustrations, of course, but does not feel miserable on the whole.

The first objection founded on this case is based, not just on the claim that the parents have done something wrong (which I certainly grant), but on the more specific claim that they have *wronged the child*. I maintained, in effect, in Section II that a creature has not been wronged by its creator's creating it if both of the following conditions are satisfied.[4] (4) The creature is not, on the whole, so miserable that it would be better for him if he had never existed. (5) No being who came into existence in better or happier circumstances would have been the same individual as the creature in question. If we apply an analogous principle to the parent-child relationship in Case (*A*), it would seem to follow that the retarded child has not been wronged by its parents. Condition (4) is satisfied: the child is happy rather than miserable on the whole. And condition (5) also seems to be satisfied. For the retardation in Case (*A*), as described, is not due to prenatal injury but to the genetic constitution of the child. Any normal child the parents might have conceived (indeed any normal child at all) would have had a different genetic constitution, and would therefore have been a different person, from the retarded child they actually did conceive. But – it is objected to me – we do regard the parents in Case (*A*) as having wronged the child, and therefore we cannot consistently accept the principle that I maintained in Section II.

My reply is that if conditions (4) and (5) are really satisfied the child cannot have been wronged by its parents' taking the drug and conceiving it. If we think otherwise we are being led, perhaps by our emotions, into a confusion. If the child is not worse off than if it had never existed, and if *its* never existing would have been a sure consequence of its not having been brought into existence as retarded, I do not see how *its* interests can have been injured, or *its* rights violated, by the parents' bringing it into existence as retarded.

It is easy to understand how the parents might come to feel that they had wronged the child. They might come to feel guilty (and rightly so), and the child would provide a focus for the guilt. Moreover, it would be easy, psychologically, to assimilate Case (*A*) to cases of culpability for prenatal injury, in which it is more reasonable to think of the child as having been wronged.[5] And we often think very carelessly about counterfactual personal identity, asking ourselves questions of doubtful intelligibility, such as, "What if I had been born in the Middle Ages?" It is very easy to fail to consider the objection, "But that would not have been the same person".

It is also possible that an inclination to say that the child has been wronged may be based, at least in part, on a doubt that conditions (4) and (5) are really satisfied in Case (*A*). Perhaps one is not convinced that in real life the parents could ever have a reasonable confidence that the child would be happy rather than miserable. Maybe it will be doubted that a few changes in chromosome structure, and the difference between damaged and undamaged genes, are enough to establish that the retarded child is a different person from any normal child that the couple could have had. Of course, if conditions (4) and (5) are not satisfied, the case does not constitute a counterexample to my claims in Section II. But I would not rest any of the weight of my argument on doubts about the satisfaction of the conditions in Case (*A*), because I think it is plausible to suppose that they would be satisfied in Case (*A*) or in some very similar case.

V

Even if the parents in Case (*A*) have not wronged the child, I assume that they have done something wrong. It may be asked *what* they have done wrong, or *why* their action is regarded as wrong. And these questions may give rise to an objection, not specifically to what I said in Section II, but more generally to my rejection of proposition (*P*). For it may be suggested that what is wrong about the action of the parents in Case (*A*) is that they have violated the following principle:

(*Q*) It is wrong to bring into existence, knowingly, a being less excellent than one could have brought into existence.[6]

If we accept this principle we must surely agree that it would be wrong for a creator to make a world that was less excellent than the best he could make, and therefore that a perfectly good creator would not do such a thing. In other words, (*Q*) implies (*P*).

I do not think (*Q*) is a very plausible principle. It is not difficult to think of counterexamples to it.

Case (*B*): A man breeds goldfish, thereby bringing about their existence. We do not normally think it is wrong, or even prima facie wrong, for a man to do this, even though he could equally well have brought about the existence of more excellent beings, more intelligent and capable of higher satisfactions. (He could have bred dogs or pigs, for example.) The deliberate breeding of human beings of subnormal intelligence is morally offensive; the deliberate breeding of species far less intelligent than retarded human children is not morally offensive.

Case (*C*): Suppose it has been discovered that if intending parents take a certain drug before conceiving a child, they will have a child whose abnormal genetic constitution will give it vastly superhuman intelligence and superior prospects of happiness. Other things being equal, would it be wrong for intending parents to have normal children instead to taking the drug? There may be considerable disagreement of moral judgment about this. I do not think that parents who chose to have normal children rather than take the drug would be doing anything wrong, nor that they would necessarily be manifesting any weakness or defect of moral character. Parents' choosing to have a normal rather than a superhuman child would not, at any rate, elicit the strong and universal or almost universal disapproval that would be elicited by the action of the parents in Case (*A*). Even with respect to the offspring of human beings, the principle we all confidently endorse is not that it is wrong to bring about, knowingly and voluntarily, the procreation of offspring less excellent than could have been procreated, but that it is wrong to bring about, knowingly and voluntarily, the procreation of a human offspring which is deficient by comparison with normal human beings.

Such counterexamples as these suggest that our disapproval of the action of the parents in Case (*A*) is not based on principle (*Q*), but on a less general and more plausible principle such as the following:

(*R*) It is wrong for human beings to cause, knowingly and voluntarily, the procreation of an offspring of human parents which is notably deficient, by comparison with normal human beings, in mental or physical capacity.

One who rejects (*Q*) while maintaining (*R*) might be held to face a problem of explanation. It may seem arbitrary to maintain such a specific moral principle as (*R*), unless one can explain it as based on a more general principle, such as (*Q*). I believe, however, that principle (*R*) might well be explained in something like the following way in a theological ethics in the Judeo-Christian tradition, consistently with rejection of (*Q*) and (*P*).[7]

God, in His grace, has chosen to have human beings among His creatures. In creating us He has certain intentions about the qualities and goals of human life. He has these intentions for us, not just as individuals, but as members of a community which in principle includes the whole human race. And His intentions for human beings as such extend to the offspring (if any) of human beings. Some of these intentions are to be realized by human voluntary action, and it is our duty to act in accordance with them.

It seems increasingly possible for human voluntary action to influence the genetic constitution of human offspring. The religious believer in the Judeo-Christian tradition will want to be extremely cautious about this. For he is to be thankful that we exist as the beings we are, and will be concerned lest he bring about the procreation of human offspring who would be deficient in their capacity to enter fully into the purposes that God has for human beings as such. We are not God. We are His creatures, and we belong to Him. Any offspring we have will belong to Him in a much more fundamental way than they can belong to their human parents. We have not the right to try to have as our offspring just any kind of being whose existence might on the whole be pleasant and of some value (for instance, a being of very low intelligence but highly specialized for the enjoyment of aesthetic pleasures of smell and taste). If we do intervene to affect the genetic constitution of human offspring, it must be in ways which seem likely to make them *more* able to enter fully into what we believe to be the purposes of God for human beings as such. The deliberate procreation of children deficient in mental or physical capacity would be an intervention which could hardly be expected to result in offspring more able to enter fully into God's purposes for human life. It would therefore be sinful, and inconsistent with a proper respect for the human life which God has given us.

On this view of the matter, our obligation to refrain from bringing about the procreation of deficient human offspring is rooted in our obligation to God, as His creatures, to respect His purposes for human life. In adopting this theological rationale for the acceptance of principle (*R*), one in no way commits oneself to proposition (*P*). For one does not base (*R*) on any principle to the effect that one must always try to bring into existence the most excellent things that one can. And the claim that, because of His intentions for human life, we have an obligation to God not to try to have as our offspring beings of certain sorts does not imply that it would be wrong for God to create such beings in other ways. Much less does it imply that it would be wrong for God to create a world less excellent than the best possible.

In this essay I have argued that a creator would not necessarily wrong anyone, or be less kind to anyone than a perfectly good moral agent must be, if he created a world of creatures who would not exist in the best world he could make. I have also argued that from the standpoint of Judeo-Christian religious ethics a creator's choice of a less excellent world need not be regarded as manifesting a defect of character. It could be understood in terms of his *grace*, which (in that ethics) is considered an important part of perfect goodness. In this way I think the rejection of proposition (*P*) can be seen to be congruous with the attitude of gratitude and respect for human life as God's gracious gift which is encouraged in the Judeo-Christian religious tradition. And that attitude (rather than any belief that one ought to bring into existence only the best beings one can) can be seen as a basis for the disapproval of the deliberate procreation of deficient human offspring.[8]

The University of California at Los Angeles
Los Angeles, California

NOTES

[1] What I am saying in this paper is obviously relevant to the problem of evil. But I make no claim to be offering a complete theodicy here.

[2] Leibniz held (in his *Theodicy*, pt. I, sec. 8) that if there were no best among possible worlds, a perfectly good God would have created nothing at all. But Leibniz is mistaken if he supposes that in this way God could avoid choosing an alternative less excellent than others He could have chosen. For the existence of no created world at all would surely be a less excellent state of affairs than the existence of some of the worlds that God could have created.

[3] Perhaps I can have a right to something which would not benefit me (e.g., if it has been promised to me). But if there are such non-beneficial rights, I do not see any plausible reason for supposing that a right not to be created could be among them.

[4] I am not holding that these are necessary conditions, but only that they are jointly sufficient conditions, for a creature's not being wronged by its creator's creating it. I have numbered these conditions in such a way as to avoid confusion with the numbered characteristics of worlds in Section II.

[5] It may be questioned whether even the prenatally injured child is the same person as any unimpaired child that might have been born. I am inclined to think it is the same person. At any rate there is *more* basis for regarding it as the same person as a possible normal child than there is for so regarding a child with abnormal genetic constitution.

[6] Anyone who was applying this principle to human actions would doubtless insert an "other things being equal" clause. But let us ignore that, since such a clause would presumably provide no excuse for an agent who was deciding an issue so important as what world to create.

[7] I am able to give here, of course, only a very incomplete sketch of a theological position on the issue of "biological engineering."

[8] Among the many to whom I am indebted for help in working out the thoughts contained in this paper, and for criticisms of earlier drafts of it, I must mention Marilyn McCord Adams, Richard Brandt, Eric Lerner, the members of my graduate class on theism and ethics in the fall term of 1970 at the University of Michigan, and the editors of the *Philosophical Review.*

BIBLIOGRAPHY

[1] Grua, Gaston: 1953, *Jurisprudence universelle et théodicée selon Leibniz*, Paris.
[2] Leibniz, G. W.: 1710, *Theodicy*.
[3] Plato: *Timaeus*.

WILLIAM F. MAY

PARENTING, BONDING, AND VALUING THE RETARDED

Offspring of the lower primates cling to the bodies of their mothers; marsupials hang on for dear life. But the human infant is unable to cling. The mother must carry her young. This dependency upon the mother for the simple act of carrying, the precondition of so much else, makes the human infant rely heavily on bonding. A mother must actively cradle her child, cherish it, for it to flourish.

The act of bonding engenders loyalty to the being and well-being of another. In a sense, the word 'loyalty' is too weak a term. Bonding describes the way in which two people settle into one another's bone marrow and kidneys, imagination and bowels. Bonding does not require a mystical merger of identity between two partners (more about that later), but it establishes a tie so powerful that neither can undertake much without reckoning with its consequences for the being and well-being of the other.

Most attempts to assess the value of the retarded child neglect the process of parenting and bonding. They proceed along the more impersonal lines implied in the title of the recent conference, "Natural Abilities and Perceived Worth." They abstractly consider those measurable properties (such as intelligence) that the retarded and others possess which distinguish them in value from zebras, parsnips, and pieces of granite. By 'abstract consideration' I mean the kind of assessment of natural abilites that a professional or policy-maker might make in an institutional setting, cleanly distanced from the possessor of those capacities. From this lofty perspective, parental imputations of worth seem merely subjective, lacking in valency.

Alternatively, this essay explores the kind of valuing that goes on within the context of parenting and bonding. It emphasizes what might best be called a relational rather than a possessional view of the self. It explores the relationship *between* human beings for its clue to their being and value and our obligations to them. It looks to the dynamics of bonding between parents and their retarded child for its understanding of worth and threats to worth rather than to a distanced consideration of capacities and performance.

Since no account of the moral life can dodge adversity, this narrative will also have to deal with the great obstacles to bonding – both external and institutional, internal and personal – that the parents of the retarded

L. Kopelman and J. C. Moskop (eds.), Ethics and Mental Retardation, 141–160.

face; these obstacles, moreover, throw their shadow across all parenting, whatever the abilities of the child.

I. EXTERNAL OBSTACLES TO PARENTING

Most researchers today emphasize the institutional obstacles placed in the way of parenting and bonding, especially in the crucial hours and days immediately following birth. The modern nuclear family, we are endlessly told, suffers simultaneously from the absence of those community supports of family and friends that helped parents bond to their children in traditional societies and from the overbearing presence of the hospital and the medical staff at birth.

The celebrative atmosphere of home birth in traditional societies supported the mother's growing attachment to the child. The excitement of other participants in the event imparted to her some buoyancy and confidence. By contrast, the industrialized West has tended until recently to treat birth even normal birth, "as an illness or an operation," and located it in an institutional setting that often intimidates women and makes them feel inadequate.[1] The modern medical concern to prevent infection curtailed that physical contact between mother and child so crucial to the bonding process. Most normal births in the hospital setting condemned mothers to several days of deprivation. Hospitals have handled premature and abnormal babies even more antiseptically. These babies became "monstrous" in the literal sense of that term; that is, objects to be seen and pointed at but not touched. It is symbolically fitting that the first director of a nursery for premature babies in this country, Martin Cooney, exhibited them at almost all major fairs and exhibitions in the United States from 1902–1940. Receipts from his *Kinderbrutanstalt* at the Chicago World's Fair of 1932 ranked second only to ticket sales for the fan dancer, Sally Rand – someone else who was to be seen but not touched. As late as 1970, only 30% of mothers were permitted to touch their babies in the first days of life. Cooney discovered, not surprisingly, that some of the mothers whose babies were out on exhibition did not want them back when they reached five pounds and were ready for return into their hands. Bonding had not occurred. Cooney's commercial techniques were distasteful, but his procedures significantly shaped methods of newborn care in the United States and not for premature babies only.

Recent critics have severely criticized medical professionals and institutions for their disruption of the bonding process. Perhaps some of these critics have exaggerated. They assumed that the experience of the first few hours and days after birth irreversibly stamps the relationship of parents to their child. They have promised almost magical results from the practice of laying the baby across the mother's warm abdomen immediately after birth. Still, it would seem that the critics have a point. Klaus and Kennell report that breast-feeding is harder to establish, incidents of child-battering increase, and, in the absence of early contact, some statistical differences in IQ and language attainment show up as much as two years later ([6], p. 39).

The additional health care needs of the retarded or handicapped baby compound the problem. Increased professional and institutional interventions have tended to reduce the mother's contact with the baby during the sensitive period so crucial to bonding. Psychologically, the mother and father need this contact. Every mother needs time to adjust to the appearance of her infant (often one to three days); the stark information that there is something wrong with the child heightens fears. Often seeing and touching the child – even a deformed or retarded child – makes it easier for parents to cope.

Changes in hospital practices since 1970 reflect a greater institutional respect for the more relational needs of the child and parents. T. V. spectacular medicine still dominates the intensive care unit for children, but hospitals have shown more appreciation recently for the parent's needs for information and access and for the work of handling, nurturing and sheltering their child. Professionals cannot ruthlessly suspend these activities and treat the child simply as a checked piece of luggage – containing so many grams of body weight, lab values, and deficits – and expect the child to come home and be embraced.

II. INTERNAL OBSTACLES TO PARENTING

It is doubtful whether the external, institutional obstacles to parenting and bonding would loom so large unless parents faced even more formidable internal obstacles in attaching themselves to their child. Three features of parenting present difficulties for the bonding process – whatever the institutional setting and whatever the child's status. I have in mind the way in which the infant confronts its parents as a stranger, threatens them with the restriction of their freedom, and stirs in them anxiety about the future. The retarded child pushes each of these experiences to its extreme.

(1) The Infant as Stranger. Moral reflection misses the test that individuals face in parenting and bonding unless it reckons with the experience of the child as a stranger. An aversion to the strange is one of the deepest psychic

responses within us. This aversion is as old as the day of our birth when we complained in no uncertain terms about being ejected from the comfort and warmth of the womb into a strange world. It appeared at seven months of age when we first became fearful of strangers and abandonment.[2] It continued at four years when we shied back behind our mother's skirts, as strangers leaned over, grinning, to tell us what a fine little boy or girl we were. And at ten we heard mysterious warnings not to go anywhere with strangers. They are tricky. Mother invested the stranger with an unspecified power to do harm. Nor does adult life leave the moral problem behind. Especially in this country, a country of immigrants, wave upon wave of strangers have made their way into established cities, neighborhoods, schools, unions, clubs and businesses, assaulting the psyche and provoking patterns of recoil and aversion.

This disturbing experience of the strange stirs no less at the arrival of the newborn. In a sense, birth, especially first birth, presents a young woman and man with the ultimate stranger, the newcomer, the absolute immigrant into this world. The young woman in 'How Green Was My Valley' announces her pregnancy to her former boy friend by saying, "We're going to have a little stranger." Something quintessentially new, faintly threatening, is about to come into the world. The place of birth today in an unfamiliar setting and in the hands of masked pros reinforces the feeling of trepidation before the strange. But the product itself, more than the setting in which it arrives, unsettles the parents. How so? The human imagination itself contributes to the surprise.

In anticipation of their child's birth, parents tend to imagine the arrival of a strapping, healthy baby, lustrous and responsive to maternal care. They have already seen dozens of infants in supermarket strollers, or Gerber portraits of infants wrapped around jars on food shelves, or Polaroid shots of babies that gestate before one's eyes. This imaginative anticipation tends to reduce the child-to-be to something relatively familiar, and, to that degree, manageable. The child will merely perpetuate, extend, and amplify the world that they already know rather than propel them toward the truly novel and strange. The world-to-be will contain no shock, no surprises. The present will overtake the future by way of peaceful annexation.

But the baby's arrival upsets the daydreaming imagination. The mother discovers that the infant isn't herself cloned; it doesn't even resemble the father. It's a prune, a monkey, a strange looking thing. Further, it imposes its strange movements, its jarring demands, its noises at all hours; it turns one's life around rather than adds something to it. When it cries one wants to mother it, and yet it also makes the stomach nervously coil.

If the ordinary child presents itself as a stranger, the retarded child does so shockingly. It shatters all parental expectations about their child. It converts the daydream into a nightmare. It presents the parents with a reality so alien to all their original hopes that the event confronts them with the force of bereavement. The arrival of the retarded kills the dream child and forces parents to grieve its loss [8]. This presents them with special difficulties in attaching themselves to the actual child, since the process of bonding – to appeal to the jargon of the experts in the field – is monotropic. Close attachments are best formed one child at a time. It is difficult to form a new attachment while grieving the loss of another. "We have noted in many parents who have lost one of a twin pair that they have found it difficult to mourn completely the baby who died and at the same time to feel attached to the survivor" ([6], p. 84). The death of the dream child means the end of the world as it was for the mother and father. It is difficult to take up the new world that presents itself in the child.

Not all parents, of course, discover immediately that their child is retarded. Sometimes they bond to the child first and only subsequently discover mild to severe retardation. That changes, to be sure, the experience, but it does not wholly eliminate the trauma of coping with the element of the strange beneath the already accepted face.

Parenting, bonding, and valuing the retarded must reckon with the fact of strangeness and all the aversion which it entails. The Enlightenment notion of the unity of humankind, based on the possession of common properties, offers little help in reckoning with the strange. The Enlightenment ideals of tolerance and benevolence attempted to trivialize differences by making them measurable on a scale of capacities and standards. As such, they underestimated the stranger who upsets my universe not because he differs from me in measurable ways but because his very existence upsets my standards of measurement altogether. The stranger is a hemorrhage in my universe.

Bonding then is no cozy process of assimilation or mystical merging of two beings. Like the love that precedes it (heterosexual love means, literally, love of the strange sex), parenting, bonding, and valuing mean openness to the strange, learning devotion to the other in the midst of strangeness. Nor does bonding, when it is healthiest, eliminate the element of strangeness. Martin Buber shrewdly observed that "othering" goes on in the I-Thou relationship. A relationship weakens when two beings take one another wholly for granted or presume too much on likeness. They no longer need to watch, listen or attend to one another. The element of strangeness is the bracer in love. From the perspective of the biblical tradition, the theological warrant for openness to the stranger is not the liberal assertion that all men are to be tolerated because, underneath, they are all alike but because God himself is the primordial stranger who bonds to his creatures despite the abyss between them.

(2) Parenting means not only coping with the strange, but also losing a portion of one's freedom. Grief over that particular loss used to center in the institution of marriage itself. But less so today. Both husband and wife today work, allowing each to continue in the life style that bachelorhood permits. Having a child, more than marriage, makes the difference. It means the loss not only of money (double salary), but of time. The mother feels drawn to love the child, to care for it, to delight in it, but it also ties her down, takes her time. The baby turns out not to be a toy; it eats; it sleeps (at the beginning, 14–18 hours per day); it eliminates (at the outset, it urinates 18 times a day, it defecates 4–7 times a day, and takes at least 8 diaper changes).

Parenting, in brief, poses the moral/metaphysical problem that Huckleberry Finn faced. The novel darkly suggests that love and freedom are incompatible. W. H. Auden once observed that he could not understand how Americans could consider Mark Twain's novel a children's book. Huck Finn runs away from his stifling home-town to get free. The river, for him, means liberty. But Huck meets up with Jim and gets bonded to him. His feeling for Jim makes him uneasy; it ties him down. Mark Twain presents a gloomy vision; love and freedom diminish one another. Just so, today, young people get out of the house and become sexually active and it tastes like freedom; but the exercise of this freedom one day produces the event that suddenly and relentlessly ties them down. The growing bonds of love seem like bondage.

The retarded child intensifies this experience of loss, certainly for those who keep the child home. Parenting inevitably means a curtailment of liberty but, ordinarily, parents can view this loss as temporary. When their children grow up, parents hope to regain what they have lost. But the retarded child is a permanent, not just a temporary, dependency. Only the death of the parents – or the child – will bring an end to their responsibilities.

(3) The act of parenting expresses hope, but, at the same time, it stirs up anxiety about the future. Clearly, giving birth affirms the future. From the ancient Manichaeans forward, pessimists have always resisted having children. The Manichaeans loathed the material world, the body, sex, and the perpetuation of that world through progeny. The trappings of their mythology have fallen away, but the basic links still persist to this day

between pessimism about the future and a great reluctance to have children: "the world is already overcrowded," "I wouldn't want to go through that again," "it would be wrong to bring children into the world."

A willingness to have children strikes a somewhat more optimistic note, but, at the same time, having a child also stirs up anxiety about what lies ahead. The affirmation contained in a child thrusts parents out into the unknown on behalf of the precious but powerless. In ordinary parenting, apprehension about the future tends to subside as parents release themselves into the round of daily demands and joys. But the fear of the future continues to hang over the parents of the retarded. In the first stage, parents often suspect something is wrong, but they know not what. It has not yet been named, rendered specific. (Fear is most intense, argued the existentialists, before it is rendered concrete.) Then, physicians name the problem. To that degree, it acquires some specificity, but ordinarily its full ramifications are not yet apparent. As the dreadful implications and complications of the original diagnosis unfold, parents begin to generalize about the future: whatever it is, it will be bad. Therefore, defensively, they turn to the present and do what they can – one day at a time. Sufficient unto the day is the evil thereof. In one important respect, however, fear of the future does not recede for parents of the retarded; the long range future still disturbs. It threatens especially at those moments when parents have occasion to see retarded adults. As much as they feel that they have come to accept their retarded child and come to terms with their own responsibilities, they are disturbed to discover that they have not yet found it within themselves to accept the adolescent and the adult-to-be. Helen Featherstone describes a visit to see older children. It posed for her as a parent all over again the confrontation with the strange. "The children . . . had lost their otherworldly charm; they were now simply psychotic adolescents. The hopes of parents and teachers had faded along with their baby fat" ([2], p. 25). But even if one accepts the future adult-to-be, the bleak circumstances which that adult will face cannot help but disconcert. "Most of us, for all our hopes and dreams, are still fattening up our children for the inevitable institutional kill" ([3], p. 180).

III. MOMENTS IN BONDING

I should concede at this point a limitation in this account of the moments in the process of bonding. This essay deals with bonding in one direction only – from parents to retarded child, not from child to parents. As such, it falls short of a full treatment of the subject. Bonding is a two-way street; parents and child interact. A baby that self-confidently attaches itself to its mother is likely to behave in ways that arouse more endearing maternal responses. Conversely, a child who has suffered traumatic institutional separation from its mother is more likely to react with anger, anxious possessiveness, or self-protective detachment. This obstructive behavior, in turn, is more apt to produce negative responses from the mother that increase the child's anger, possessive clinging, or guarded apathy ([1], p. 340). Further, the varying degrees of mental retardation, from profound to mild, make a difference in the obstacles parents face in bonding. A full-dress treatment of bonding would have to take account of these differences.

Recent efforts to describe the awesome process of the bonding to the retarded child sound very much like Kübler-Ross on death and dying; the responses of anger, fear, denial, bargaining, and acceptance surface in the discussion. This approach has the advantage of letting the actual emotional experience of parents see the light of day. As in other crises to which her stages have been applied, however, one wonders whether Kübler-Ross and those who follow her interpret these stages as descriptive or normative, chronological or concurrent.

An alternative scheme for interpreting the process of bonding focuses less on specific emotional attitudes than on the magnitude of birth and the changes it calls for in parents' lives. Having a retarded child imposes upon parents an experience that corresponds structurally to those great turning points (and associated rites of passage) in traditional societies that transported people from one stage of life to another – birth, puberty, marriage, and death. These turning points included three "moments" or elements: (1) detachment from the past; (2) a period of transition; and (3) attachment to a new life and estate [9, 10]. Whatever the experience of having a retarded child may entail emotionally for parents – fear, loneliness, resentment, and the like – it includes three "moments." It wrenches them out of their former life, its assumptions and priorities; it produces a period of disorientation, confusion and provisional adjustment; and it waits upon the parents' final attachment both to the child and to the circumstances that life with that child entails. In these respects, it resembles ancient rites of induction into the sacred. In the absence, however, of definitive social rites today of the kind that obtained in traditional societies to shape and limit great crises and turning points, these "moments" are not neatly successive. They are distinguishable but overlapping in the course of a profound sea change in the life and prospects of the parents.

Detachment from the Past

Detachment, at times, seems too mild a word: destruction is closer to the truth of the matter. The birth of a retarded child shatters one's world. "The world crumbled and fell around us," reports the mother of a Down's syndrome child ([2], p. 220). "In one cataclysmic moment, our world had been shattered," another mother testifies ([2], p. 220). Helen Featherstone, who has gathered the evidence but also writes out of personal experience, avers, "The shock of disability seems to obliterate the life that exists," and ". . . threatens to define an ominous pattern for the new one" ([2], p. 220).

The earlier discussion of coping with the strange emphasizes this experience of break with one's previous life. Routines, demands upon oneself and others, daily and long-range prospects change. The retarded child loosens one's assumptions about what is important and unimportant. The child is a stranger in the sense that it estranges parents from their familiar world. Things look different – not only in the way parents view their own lives, but also the lives of their friends. Parents testify to their sense of isolation. This isolation shows up at practical levels, but it also bespeaks a deeper cultural isolation.

At the practical level, isolation begins, of course, in the hospital where the medical interventions often separate the mother not only from the child but other mothers. It spreads as friends and relatives do not know how to acknowledge the birth. (One mother reports that nothing unnerved her so much as her own mother's insistence that she institutionalize her child. Irony abounds in the advice. In offering it, the grandmother was attempting to mother her own daughter – protect her from suffering. But the daughter found the advice shattering. It undermined her own ability to mother her retarded child. An important feature of her capacity to mother her child derived from the mothering she had received in childhood. But now her own mother seemed to be saying: "If you had been retarded, I would not have mothered you." The daughter kept her child, but it disturbed other bonds.) In the course of time, the child's special needs for care necessarily affect the family's daily routine and social life. Brothers and sisters sometimes feel cut off from their peers by their heavier responsibilities. One mother testifies that unhappiness is alienating not only because others shun the unhappy but because the unhappy resent the other's happiness – especially when the fortunate take for granted what they have.

At a cultural level, the birth of a retarded child cashiers not only the parents' social life, but also their assumptions about prized American values:

the American reliance on the wizardry of medicine, its orientation to youth as a symbol of achievement and promise, and its abhorrence of dependency. Americans have been given to a technological triumphalism – the assumption that any and all human problems will eventually yield to the organized assault of research and the strategic weapons generated by that research. But the parents quickly discover that medical research and the experts who wield the technologies which flow from it quickly reach their limits in dealing with this child. A whole army comes to a halt and becomes an irrelevancy before their own tiny baby and its fateful problem. The physician cannot function as an heroic interventionist, merely as the messenger of bad news. Medical magic that prolongs their child's life or patches up its other physical difficulties merely highlight medicine's inability to touch the fundamental cause of their grief.

Further, the event undercuts the conventional American attitude toward youth. As an immigrant people, Americans orient themselves more obsessively toward their children's performance than members of more settled societies. In America, immigrant children have lived under the pressure to outstrip their parents; in Europe, merely to live up to them. The parents of the retarded, however, find it impossible for the child to meet either demand. The child's severe impairment both removes and yet fails to remove them from the myths about youth. Parents suffer the discrepancy between myth and performance by feeling guilty and responsible for the child's shortfall. The American middle-class myth of producing the perfect child has always tended to confuse the process of manufacturing with the act of giving birth. One takes credit for the child as a product rather than receives it as a gift. In this atmosphere, the arrival of a retarded child signifies a personal failure rather than an impersonal blow of fate. The child excommunicates its parents from the ordinary circle of nursery achievement. A child becomes "a dispiriting symbol of shared failure" ([2], p. 91) between husband and wife. Guilt and self-doubt grow apace and compound the problem of attachment. As long as the child is a sign of failure, it is hard to love, hard to attach to, because, in accepting the child, the parent must ultimately learn to accept his or her discredited self.

In yet another way, parents of the retarded feel isolated from the prevailing culture. Americans highly prize independence both in their national and personal life. The dark side of this character trait is their abhorrence of dependency. Americans do not like to depend upon anyone personally and they tend to withdraw from those who threaten to depend, in turn, upon them. That is why Europeans so often complain that Americans, beneath their surface friendliness, have little capacity for friendship and why recent migrants to the mainland from Puerto Rico have found more than the weather cold in New England. Whether or not independence deserves praise as a character trait, the failure to achieve it either as a family or in any of its members, stigmatizes and isolates the family from others. The irremediable dependency of their severely or profoundly retarded child throws the family into a social plight that most others in society abhor and shun.

In a variety of ways, the retarded child disconnects its family from the culture at large. It is un-American; it makes aliens of its parents, brothers and sisters. For better or for worse, they cannot participate umbilically in the technological momentum, the orientation to youth, and the pride in independence that so mark their peers. They become a little estranged in their own culture. To become loyal to this child requires a reconsidered relation to much else.

Transition

In traditional societies, the passage from one estate to another (birth, puberty, marriage, war, peace, death) is a particularly delicate and perilous time. It often entails a suspension of ordinary activities associated with either the life left behind or the life to be assumed. It is a potentially confusing period in which specialists (priest, warriors, and shaman) play an especially important role. They supervise and assist people through the transition rites associated with the great turning points in life.

I have already mentioned the experience of isolation and confusion that can engulf parents with the arrival of the retarded infant – removal from others and disorientation in their perception of themselves and their goals. Professionals can err in opposite ways in their handling of this transition period. Out of compassion, caution, or their own aversion to pain, they may unduly prolong the transition period by withholding the worst news. They dole it out to the parents in bits and pieces. Sometimes such compassion misses the mark. In the absence of candor, parents often generate still worse fears. Anxiety feeds on uncertainty. Indeed, it is impossible to bring the transition period itself to an end if parents fear that the future may hold even worse perils not yet revealed.

Professionals, at the other extreme, may fail sufficiently to respect the need for a transition period. They may be unduly impatient with parents and unwilling to give adequate time for the assimilation of bad news. One doctor confesses, "It isn't okay to break the news, and support them for a week, and then give them an infant stimulation program and everything is fine. Some parents turn around very fast, and some turn around slowly . . ." ([2], p. 188). They assume that parents should absorb the news and get on with the regimen. They do not realize that this kind of news both shocks and burns its way slowly into the mind.

Some conversion time is required (conversion in the sense of the original meaning of the Greek term metanoia – to turn around) for parents to re-orient themselves. Freud sensed the need for this turn-around time in any profound reconstructions of the self. When asked why the analyst could not simply tell the patient the truth about himself and be done with it, Freud remarked that the truth so delivered would not be assimilated; it would sit like a parallel deposit in the mind.

Just so, parents of the retarded need a period of assimilation. They need some time to absorb the news that severs them from their former expectations, myths, and gratifications. They also need some time to enter into their new estate. The child (and their life with the child) has its values, but the values strike with a force that transvalues rather than merely adds to the values which they already profess.

Attachment

The process of healthy attachment includes a number of prenatal and immediate postnatal experiences that support bonding: (a) prior to pregnancy, at least some contact with babies (even if only babysitting); (b) the experience of quickening during pregnancy (the perception of the fetus as a separate individual sometimes converts unwanted to more acceptable or wanted pregnancies); (c) birth (studies suggest that those who witness birth are more likely to bond to the child than those who do not); (d) immediate contact after birth (mothers deprived of access to the baby during the so-called sensitive period, are inclined to be less attached; the baby "belongs," as it were, to someone else); (e) a favorable institutional setting (homelike, if not at home, with professional staff assisting, rather than displacing the mother); (f) reciprocal interaction (touch, eye-to-eye contact – the child can see at birth), oral contact (the neonate prefers the high-pitched female voice), entrainment (a type of dance-movement in the course of conversation between mother and child), odor (by the fifth day, a baby can tell its mother's breast pad), heat, and heartbeat (to say nothing of bacterial nasal flora and antibodies imparted with the mother's milk) ([6], pp. 86–94).

The proposal that final attachment to the retarded child resembles induction into the sacred in traditional societies can be interpreted sentimentally or realistically. Sentimental testimony abounds. One parent professes, "I believe a brain-injured child is a gift from God. This child is a precious gift to a family" ([2], p. 218). Another parent avers, "The child we have is like an angel. Treat him well, he will put in a good word for us with God" ([2], p. 166). Medieval people sometimes referred to the retarded as "infants of the good God" ([7], p. 7) – with special freedom to roam. The fool, we know, enjoyed a measure of liberty in the court denied to others. This special elevation of the retarded reflects itself in the very word sometimes used to designate them. The word 'cretin' has now fallen into disuse; it smacks of the pejorative. But some suggest that cretin derives from the word, 'Christian'. In effect, the retarded alone seemed to deserve the adjective so often loosely (and wrongly) used to suggest the innocent. Owing to their simplicity of mind, the retarded alone seemed incorruptible.

Such views verge on the sentimental. The phrase, 'gift of God', sometimes functions less as a theological assertion than a claim about the inevitably good consequences that follow from having a retarded child. It claims too much for the retarded's tonic impact on their families and society. It smacks too much, albeit in a poignant way, of the fervent rhetoric that one sometimes hears from AA and other support groups that need missionary zeal in order to sustain courage. Gifts from God – some surely may seem, but not inevitably so received or perceived. Taking responsibility for the retarded may deepen the life of some family members. But suffering does not inevitably ennoble. Heavy responsibilities crush as well as enlarge their bearers.

Yet bonding does take place and of an order and depth that sometimes suggests a powerful transformation of life. Another parent resorts to religious language. "A child like this is a sacrament." The parent thereby concedes that having an abnormal child makes one feel "outcast, doomed somehow not to be normal oneself," but acknowledges, at the same time, "an entirely different covenant with existence" ([2], pp. 213–214). In another instance, the father of an autistic child rages against his lot and yet finally concedes, "Consciousness is like a free gift arising out of a bond that is tragic and good. But for him, I guess, and for us, the bond is what comes first" ([2], p. 237).

Not that all is well. Words such as 'tragic, doomed, outcast' still appear in the testimony. But the harshness of the experience no longer encompasses. Attachment has taken place. Featherstone wryly acknowledges a negative sign of this bonding. "At the minimum and wintry level, acceptance may begin to bring the recognition that the child's hardship no longer represents the only obstacle to happiness" ([2], p. 220). Somewhat more sunnily, the brother of a mildly retarded busboy reports taking stock of his deficient sibling. A stranger innocently asks the rather conventional question – What does your brother do for a living? The young man, for the first time, does not answer preemptively by explaining that his brother is retarded and therefore only a menial in a restaurant. "Suddenly it struck me for the first time in the twenty-eight years of our lives together that mental slowness is not the sum of [my brother's] existence. He is a man who is married, holds down a job, pays taxes." So he returns to the question and simply tells the stranger, "He's in the restaurant business!" ([2], p. 176).

It is fashionable in the literature on the retarded child to close with a discussion of "acceptance." In my own judgment, acceptance does not describe the bonded parent without qualification. Attachment, at its deepest level, sets up a tension between acceptance of the child as it is and a quest for interventions that will improve its lot. Any human being bonded to the being and well-being of another does what he can and enlists the aid of others to serve its good. Simple acceptance alone hardly describes the parents' relationship to their child as they attempt what they can to foster their child's excellence and well-being. To that end, of course, they seek the services of two interventionists – the doctor first but eventually the teacher – to stretch the child's capacities and to help it live up to them.

However, an activist compulsion alone can turn demonic; hammering the child with interventions can mean a rejection of the child as it is. It may also betray a profound self-rejection. The parent feels guilty about having produced a child with defects and attempts to atone by becoming a superparent. Sooner or later, the parent must reckon with and accept the child's limits. Otherwise, the effort to serve the child's excellence and well-being deteriorates into a battering and self-battering that hides beneath the fair face of love.

Both poles in the tension are necessary. Religiously put: attachment becomes too quietistic if it slackens into mere acceptance of the child as it is. Love must will the well-being and not merely the being of the other. But, even worse, attachment lapses into a gnostic revulsion against the given world if in the name of well-being it recoils from the child in its deprived state of being. One observer perceives the right balance between acceptance and intervention when he observes, "I admire the purpose and ingenuity with which parents and children forge a good life out of imperfect materials" ([2], p. 10). It includes efforts at transformation, yes; but not as though all value rests in the transformation, as though the imperfect materials themselves had to be surpassed for anything good to come of the life.

IV. VALUING

The problem of valuing the retarded child is misunderstood if placed wholly in a marketplace setting. I mean thereby a setting in which the retarded child appears to be a discrete item in the world to be compared and weighed against other discrete items. In this case, one asks what correlations appear between natural ability and perceived worth and then one tests out that worth on the basis of how many social resources (public and familial) one is willing to spend on behalf of the support and enhancement of the retarded. What price, in effect, can they command within the public and domestic economy? This question, so formulated, begs for an answer in the coin of utility.

This approach misses the issue in valuing not so much because it is too cruel and ruthless (though it is) but because it misses the even more cruel question which the retarded themselves pose. The retarded do not chiefly pose the question: What is the relative worth, or lack of it, of a child with my particular disabilities? But rather: What worth has a world that includes me in it? The latter is the more encompassing, metaphysical question that parents face. This child which is my child, not the child I hoped for, named, and registered for college in fancy, depresses for me utterly the world's worth. Herein lies the deep quarrel with God, the terrible envy and resentment of friends. The problem is not one of relative good in a marketplace of goods and preferences, but the problem of evil. The dead child, the child of fancy that one needs to bury with the arrival of the deprived child, raises the question of the worth of the world and the worthwhileness of life. The question deepens and darkens beyond the perceived worth of the child, relative to other goods, to the worth of anything whatsoever. Nothing seems worth a candle.

Parents see the problem at this more metaphysical level because they suffer at this level. And, at least some of them, it would appear, have sought to respond at this level when they have confessed themselves to feel outcast and doomed and yet have reckoned with the child as establishing their "covenant with existence." Bonding to a child who renders questionable the value of existence reestablishes one's covenant with existence. Neither policy makers nor professionals should deal with that testimony lightly.

This covenant with existence demands devotion beyond the ordinary measure. Featherstone likens the plight of parents to the person who happens upon a drowning man. As the only person there, she has the choice of jumping in and trying to save him or agonizing on shore. "In the first case I am a hero; in the second, a coward. There is no way I can remain what I was before – an ordinary person" ([2], p. 83). But it would be a mistake to claim that heroism alone sustains the covenant, as though parents receive no sustenance from the child. That sustenance may be difficult to acknowledge without sounding as though one is justifying the existence of mental retardation. Yet some parents have acknowledged the deepening of their lives; they find themselves in retrospect a little kinder, a little gentler, a little more sensitive to the difficulties others face than they might otherwise have been.

If this is the case, then valuing the retarded is not simply a matter of placing a price tag on a product that someone has manufactured, as though values themselves are invented, fabricated, and marketed; they are also discovered and honored; they impart strength to those who do the valuing. Values are not ideal demands alone; they are also statements about strength and power.

Some of this lurks in the Latin root for value – valere – from which we derive not only value, but valor and valid and convalescent. The convalescent is someone whom illness has assaulted but who begins to grow again in strength. Similarly, the valorous man or woman is strong, stout, brave, vigorous, and powerful in the face of adversity. The brave person exudes power in the face of odds.

Similarly, when we refer to values – to the value of the family, work, love, birth, and the retarded child – we are not talking about frothy matters that we imagine or invent, we are talking about people, institutions, ideals and demands that bear down upon us; they validate themselves, as it were, as stout, strong, vigorous, and powerful, prevailing upon us as worthy of response in our freedom. Values elicit commitments from which, in turn, strength and power derive.

V. AN EPILOGUE ON POLICIES AND QUANDARIES

This essay has tried to put the problem of valuing the retarded in a relational setting. It has looked to the relationship between parents and child for its clue to moral claims and obligations. It has attended specifically to the process of bonding, as it honors and values another and as it generates loyalty and trust between those bonded. The path to bonding does not usually run smooth. Parents undergo an ordeal that beggars description except in the language reserved for induction into the sacred. The testimony of parents reminds one of the language that one associates with the relational thinkers.

Kierkegaard spoke of fear and trembling in the God-relationship [5] ; Heidegger reckoned with dread in the relationship to death;[3] Buber recognized that the 'I' and the 'Thou' in the I-Thou relationship remain strange to one another in the very midst of I-Thou address. Relationship to the retarded at the deepest level requires a transformation that upsets, winnows, confuses, and turns around the self.

The use of this religious language also implies that the process of bonding has more than psychological import; its significance is ontological and moral; bonding gives a clue to the being and well-being of others, to one's obligation to them, and to one's own being and value in so honoring the tie.

This general approach departs from the prevailing tendency among moralists to resort exclusively to the notion of abstract rights and duties in dealing with the moral warrants for care of the retarded. Ethicists usually begin with natural capacities (such as intelligence) abstracted from the historical accidents of family, religion, education, and the like and seek to determine what duties we have to all human beings whatsoever who possess these capacities. This prevailing approach has led, across the last century and a half, to the establishment of huge facilities designed to care for people who, by virtue of natural misfortune or historical accident, suffer abuse or neglect at the hands of their fellows or who need care of a kind which neither they nor their families nor their friends can offer. The support for these institutions – hospitals, mental hospitals, nursing homes, and workshops – comes largely from strangers (through taxation or philanthropic giving); further, they are manned by strangers whose job it is to extend help to strangers (to all persons whatsoever, irrespective of ties).

The approach taken in this essay differs (but only partly) from this prevailing view of rights and duties and its institutional results. The generation of a set of rights and duties to the retarded on the basis of an abstract consideration of natural abilities leads to moral minimalism and to a minimal discharge of those duties in huge institutions. Duties based on rights tend to reduce to what the socially undefined person can claim against anybody. Additional obligations that grow from bonding get written off as the merely optional and preferential. They do not enter into discussion of what the society owes creatures whose being and worth disclose themselves chiefly in the context of ties. The disquieting results of this minimalism are all about us in a post-Enlightenment society, rootless and libertarian. We have tended to reduce our direct obligations to the marginal and the deviant, and to discharge what obligations we acknowledge through the device of segregating institutions and isolating services. The retarded's right to these services and the shape we have given to them results from a bondless consideration of natural abilities and worth. Meanwhile, the very scale and structure of the institutions themselves have made it more difficult for bonding to occur within their walls.

A more relational approach does not, however, argue for the dismantling of large scale institutions. Criticism ought not obscure either the philanthropic impulse that often prompted their founding or the conscientious efforts of many of their managers and staff to calibrate their limited resources to levels of need. Moreover, some versions of mainstreaming today set the needy free from segregated institutions only to abandon them to the *de facto* segregation of third-rate, big city hotels. Care deteriorates from little to nothing.

At the policy level, a respect for the importance of bonding should encourage the reform rather than the dissolution of total institutions. It should attend to those various factors that foster and nurture small-scale community even in large institutional settings – architectural design, the layout of interior space, equipment, public and private rooms, and the strategic use of professional teams and volunteer groups. Further, it should also provide more adequate public supports to families or surrogate family units to ease the burden of care – more adequate training programs, respite houses that provide families with periodic relief from total care, and the mobilization of religious and other volunteer groups to supplement the family nucleus with some of those resources previously available only through the extended family.

A more relational approach hardly solves all problems, moral and practical. Clearly, it will not solve all quandaries parents face: whether to institutionalize a child, what goods will best serve the well-being of a child, what balance should be struck between the claims of the retarded child against those of other siblings, work, and marriage. Indeed, the existence of bonds often intensifies rather than solves problems in casuistry. The bonded agent must reckon with conflicting loyalties, not just competing preferences, when a decision must be made. He cannot, moreover, blunt the moral ache of the decision by resort to the marketplace device of calculating trade-offs or balancing goods over evils. He must honor the tragic, or, as religious people may prefer, the comedic element in a moral decision. A decision to institutionalize is not just the debit side to a balance. One persists in wishing otherwise, even when one chooses one's course. The brokenness, the incompleteness, the tragedy, the poignancy of a decision carries forward with us into the future – unless an ultimate perspective with more depth than a balance sheet lets us see things whole.

Finally, a relational approach cannot seriously claim to derive notions of worth wholly from the faulty testimony of those who are bonded to one another. If human beings have a worth derived from a human tie alone, the retarded child is wholly hostage to the erratic valuations of parents and caretakers, good, bad, or indifferent. That is why Enlightenment moralists tried to establish a notion of worth irrespective of ties. Their minimalism sought to protect against worse fates. Enough has been said here about the external and internal obstacles to bonding, about the awesome process of detachment, transition and attachment, to give one pause about placing authority for making ultimate judgments of worth in the hands of parents and others who face, many times unsuccessfully, that difficult ordeal. The battered child silences foolish sentimentality about the success-rate of the bonding process. If value and the perception of value derive entirely from a relationship, then the powerless one within the relationship becomes wholly hostage to it. Parents cannot be legitimated as the sole or final arbiters of the value of their children. They are frail enough; and the process by which they bond to their children is strewn with enough obstacles. This consideration and caution partly led the theological tradition, when speaking relationally, to posit a God-relationship as encompassing, authorizing, and judging all others. To testify, however, to that relationship with bonded conviction, one must suffer an induction into the sacred of which the ordeals I have described in this essay are but a petty parable and sign.

Georgetown University
Washington, D.C.

NOTES

[1] For this and following material on bonding, see [6], p. 39.
[2] Dr. Jerome Kagan has reported clinical evidence of these fears appearing at this age.
[3] Heidegger elevated the phrase 'being towards death' (*Sein zum Tode*) into one of his fundamental existentialist categories in his masterwork, *Sein und Zeit* [4].

BIBLIOGRAPHY

[1] Bowlby, J.: 1969, *Attachment and Loss*, Vol. 1, Basic Books, New York.
[2] Featherstone, H.: 1980, *A Difference in the Family*, Basic Books, New York.
[3] Greenberg, J.: 1972, *A Child Called Noah*, Holt, Rinehart, and Winston, New York.
[4] Heidegger, M.: 1957, *Sein und Zeit*, Max Niemeyer Verlag, Tübingen.

[5] Kierkegaard, S.: 1954, *Fear and Trembling*, W. Lowrie (transl.), Doubleday and Co., New York.

[6] Klaus, A. and Kennell, J. H.: 1976, *Maternal-Infant Bonding*, C. V. Mosby, St. Louis.

[7] Kramer, L.: 1964, *A History of the Care and Study of the Mentally Retarded*, Charles C. Thomas, Springfield, Illinois.

[8] Solnit, A. N. and Stark, M. H.: 1961, 'Mourning and the Birth of a Defective Child', *Psychoanalytic Study of the Child* 16, 523–537.

[9] Van Der Leeuw, C.: 1963, *Religion in Essence and Manifestation*, Harper and Row, New York.

[10] Van Gennep, A.: 1960, *Rites of Passage*, University of Chicago Press, Chicago.

JOHN C. MOSKOP

RESPONSIBILITY FOR THE RETARDED: TWO THEOLOGICAL VIEWS

Like their philosophical colleagues, Christian moral theologians have written relatively little about the rights of retarded persons or about our responsibilities to them. It appears that the existence of retarded persons raises difficulties for certain theological as well as philosophical approaches to ethics. Nevertheless, two contemporary American Protestant theologians, Joseph Fletcher and Stanley Hauerwas, have directly addressed the issue of our responsibilities regarding those with different natural endowments. In this paper, I will first review and evaluate the positions adopted by Fletcher and Hauerwas. We will see that despite a common Christian heritage, their approaches stand in striking contrast to one another. I will then attempt to suggest an acceptable middle ground between the two positions.

I. FLETCHER ON PREVENTING RETARDATION

Joseph Fletcher's comments on the retarded are contained in his book *The Ethics of Genetic Control* [5], where he urges the adoption of a "Promethean" attitude toward new scientific and technological advances in genetics and reproductive biology. That is, Fletcher advocates the bold use of new technologies to control population, prevent or cure genetic disease, and arrest what he and others view as the deterioration of the human gene pool.

In order to accomplish these goals, Fletcher maintains that individuals should make both the number and the nature of their offspring matters of rational choice. The following statement is representative: "Our moral obligation is to control the quality as well as the quantity of the children we bear" ([5], p. 151). This can be done, first, by gaining greater information about possible reproductive outcomes through genetic screening and amniocentesis. Based on this information, greater control over the number and kind of children one will bear can be exercised through contraception, abortion, and use of donor sperm and ova. In addition to these already developed techniques, Fletcher looks forward to the development of clonal reproduction, artificial wombs, and direct intervention to correct genetic defects *in utero* ([5], pp. 40–41, 56).

Fletcher's evaluation of these techniques is based on his own well-known

L. Kopelman and J. C. Moskop (eds.), Ethics and Mental Retardation, 161–175.

approach to moral reasoning, "situation ethics" ([4], [5], p. 30). Situation ethics recognizes only one categorical obligation: to obey the Christian law of love. The moral agent is expected to determine the most loving course of action in each situation. Traditional moral rules such as the Ten Commandments are no longer viewed as strict constraints on action; rather, Fletcher contrasts situation ethics to what he calls 'rule ethics'. Rule ethics, an approach he attributes to more traditional theologies, holds that some things, like truthtelling, are absolutely and always right, and others, like abortion, are absolutely and always wrong ([5], pp. 30–31, 118–120).

Like John Stuart Mill, however, Fletcher interprets the Christian law of love in terms of the utilitarian principle that one ought to act so as to produce the greatest happiness for the greatest number ([4], pp. 95–99; [5], pp. 138–139). Thus, Fletcher's approach requires attention to the consequences of one's action in order to determine its probable costs and benefits. That action is best which provides the most favorable ratio of benefits to costs ([5], pp. 158–160).

What, then, is the loving, rational, good-maximizing choice regarding the "quality" of children we ought to bear? Fletcher's answer is that we should choose healthy, nonretarded children and avoid conceiving and bearing those with illness or handicap. "Controlling the quality of life", Fletcher claims, "is not negative; it just rejects what fails to come up to a positive standard" ([5], p. 158). Although this standard is never made explicit, Tay-Sachs disease and Down's syndrome are elsewhere cited as examples of "terrible diagnoses" which should be prevented through abortion ([5], pp. 48–49). Abortion, incidentally, raises no special moral problem for Fletcher, since he holds that the fetus becomes a person only at birth ([5], p. 139). Even after birth and presumed personhood, Fletcher advocates passive euthanasia for at least some newborns with serious genetic diseases ([5], p. 29, [6], pp. 140–147).

In short, the strong emphasis of Fletcher's approach is on preventing the conception and birth of retarded and handicapped children. Relatively little attention is paid to the kind of care such children should receive. Although Fletcher allows that handicapped children should become objects of our loving concern ([5], p. 152), his clear preference for preventive measures is obvious in statements like the following: "It is cruel and insane to deprive normal but disadvantaged children of the care we could give them with the $1,500,000,000 we spend in public costs for preventable retardates" ([5], p. 160). Preventing the birth of handicapped children is, in Fletcher's view, of such vital importance to the general social welfare that he favors pursuing it through "compulsory controls." Among the societal controls Fletcher favors are mass genetic screening, establishment of genetic registries, compulsory testing of prospective mates for genetic compatibility, and prohibition of marriage of couples who are at risk for genetically diseased offspring unless one or both submit to sterilization ([5], p. 179–183).

II. WEAKNESSES OF FLETCHER'S APPROACH

Fletcher certainly offers a spirited defense of a responsibility to prevent genetic disease by means of new reproductive technologies. Nevertheless, his views arouse suspicion on a number of grounds. Let us consider a few such grounds for suspicion.

First of all, Fletcher shows an uncritical enthusiasm, less about religion than about new applications for medical science and technology. For example, he envisions legitimate uses for almost every conceivable technological advancement, including cloning, man-animal, and even man-machine hybrids ([5], pp. 154–156, 171–173). Little or no mention is made of moral problems in the development of such techniques and in determining the moral status of quasi-human hybrids. Fletcher also offers a broad defense of risky fetal research on the grounds that although such research may injure the fetus, the fetus can also be injured during *natural* gestation ([5], p. 94). This however, is irrelevant; what must be justified is why the fetus should be subjected to *extra* risks in research, risks which are neither a part of that fetus' natural gestational process nor necessary for therapeutic reasons.

While his enthusiasm for new uses of medical science and technology is clear, Fletcher's debt to the tradition of Christian ethics is much less obvious. Although situation ethics was introduced as an approach based on the divine commandment of love, it owes more to a secular moral tradition, utilitarianism, than to any tenets of Christian faith. In his book *Situation Ethics*, Fletcher speaks only of a "coalition" with utilitarianism ([4], p. 95), but his later books appear to *identify* right or loving action with promoting the greatest good for the greatest number. Even though such an identification is far from self-evident, Fletcher provides no argument to support it. Neither does he take account of the serious criticisms philosophers have brought against utilitarianism.[1]

Fletcher makes no secret of his distaste for more conservative moral doctrines, particularly those of the Catholic tradition ([5], pp. 97–99, 119–120). His writing, however, inadequately characterizes opposing views, lumping all of them under the rubric of 'rule ethics'. Rule ethics, portrayed as submission to specific and absolute moral rules established by divine fiat ([5], pp. 119–120), looks suspiciously like a straw man. This description would seem to fit only the most rigid of moral theologies. It fails to do justice to the creative, intellectual activity of *developing* and *interpreting* moral principles on the basis of fundamental beliefs and values.

Even if Fletcher's utilitarian approach is granted, serious doubts can be raised about his analysis of several of the specific moral problems he addresses. For example, Marc Lappé argues that Fletcher's concern about the genetic deterioration of the species is a red herring [10]. According to Lappé, "the consensus of the best medical and genetic opinion is that whatever genetic deterioration is occurring as a result of decreased natural selection is so slow as to be insignificant when contrasted to 'environmental' changes, including those produced by medical innovation" [10, p. 360]. Rather than a burden, genetic diversity may be viewed as an evolutionary value, since it permits adaptation to future environmental demands. If, however, the genetic future of humanity is not as dark as Fletcher contends, his proposals cannot be viewed as necessary in order to prevent widespread hardship and suffering sometime in the future.

Also problematic is Fletcher's position on abortion, summarized in the following statement:

> Ethically it is in the discretion of a woman to prevent or end any pregnancy she does not want, unless she has promised the child to a husband or lover who justifiably insists on it, or unless a clear case can be made that society has a supervening interest in its birth. (Rarely indeed would either of these limitations cut into her personal freedom.) ([5], p. 151)

Fletcher does not, however, show how this permissive position on abortion can be justified on the basis of utilitarian principles or, for that matter, on the grounds that it is the most loving course of action. In fact, it is highly unlikely that this could be done, since in at least some cases the harm done to the fetus by aborting it would appear to outweigh the costs to the woman of allowing it to be born. Fletcher does offer in support of his position the claim that the fetus is not a person until birth, but this is irrelevant to a utilitarian analysis, since fetuses can presumably be benefitted or harmed whether or not they qualify as persons. Fletcher, however, never seems to take harm to the fetus into consideration in arriving at his conclusion.

Though he does take the interests of children into consideration, Fletcher's appraisal of what counts as the benefits and costs of accepting handicapped children is also open to question. He maintains that we have an obligation to the prospective children themselves, their families, and to society, to prevent their birth ([5], p. 151). Given his utilitarian approach, Fletcher seems to be claiming that for each of these groups – children, families, and society – the costs of accepting such children outweigh the benefits. Such a claim, however, can be challenged for each group.

Consider, first, the case of the children themselves. Fletcher's comments on Down's syndrome, one of his favorite examples of a genetic disease, are a vivid case in point here. Those comments are uniformly negative – persons with Down's syndrome are referred to as "pathetic creatures" ([5], p. 28), their diagnosis is "terrible," and misery is in store for them ([5], p. 49). Given this characterization, one can understand why Fletcher might hold that their life is not worth living, but those who care for persons with Down's syndrome are likely to find Fletcher's description grossly inaccurate and to insist on the value life possesses for many of these children [3, 14].

With regard to families, there is little question that the presence of a retarded child can be a serious physical and emotional burden. There is, however, also little question that a retarded child can be a joy and a great enhancement to the lives of his or her family. One cannot, therefore, on utilitarian grounds, establish a general obligation to prevent the birth of retarded children in order to avoid harm to their families.

Among the costs to society of handicapped children, Fletcher is most concerned with the financial burdens of care and the long-term effects of deleterious genes on the human gene pool. His discussion, however, neglects some other significant societal effects. His highly unsympathetic characterization and proposal for a wide ranging and compulsory social program to prevent conditions such as Down's syndrome can only serve to reinforce stigmatizing labels, attitudes, and policies, and thus add a heavy social burden to the physical ills of present handicapped persons and of those future persons whose handicaps we will not be able to prevent. It may also tend to discourage important social virtues of care and concern for those with fewer natural abilities and advantages. In short, Fletcher's claim that, financially and genetically speaking, we cannot afford to accept the handicapped may be countered by the claim that, morally speaking, we cannot afford *not* to accept them.

III. HAUERWAS ON THE RETARDED AS GIFTS

In stark contrast to Fletcher's approach is the position taken by Stanley Hauerwas in an essay entitled "Having and Learning How to Care for Retarded

Children: Some Reflections" [8]. Hauerwas argues that the Christian faith provides powerful reasons for accepting and nurturing retarded children. Though he does not refer to Fletcher, Hauerwas takes issue with Fletcher's views that childbearing should be a matter of individual choice and that we should strive to prevent the birth of children who may be retarded.

Hauerwas approaches the question, "Why should retarded children be welcomed into the world?" by reflecting on the reasons why we have children at all. He first considers what he calls 'a common presumption' that we choose to have or not to have children ([8], p. 632). This presumption, Hauerwas claims, is the cause of several widespread misconceptions about children. First of all, if children are purely the result of our choosing to have them (i.e., if there is no special reason why we should have them), then it seems that we must bear full responsibility for their well-being, we must insure that they receive the best of everything, perhaps even to the extent of trying to make use of the best sperm and ova.[2] Second, if children are the result of our choice and are sustained by our sacrifices, then they become our product, completely dependent on us as their creators. Finally, if children are to justify our choosing them and working so hard for them, we expect them to be perfect, and we feel cheated if something goes wrong with them. These misconceptions, Hauerwas argues, make the idea of having retarded children unattractive; moreover, they have corrupted our child-rearing practices for all children ([8], pp. 632–633).

In place of the notion of choosing our children, Hauerwas offers the view that children are *gifts from God*. Christians (and Jews, Hauerwas adds parenthetically) should view children as divine, not human creations. Children are independent beings called forth by God to assume their own place in the community and in the world. We do not have children in the expectation that they be perfect, but rather as a sign of our trust in God and in his world. For Hauerwas, then, having children is a basic vocation or duty, a sign of our commitment to carrying on the Christian community in which we live and believe.

Although these beliefs form the basis of a Christian approach to all children, Hauerwas claims that they apply with special intensity to the retarded. Hauerwas agrees with Fletcher that retarded children are not what we would choose to have; for him, however, this underlines that fact that having children is not a matter of choice. Because retarded children are not our creations, their independence from us and need for separate development is apparent. Because we as parents are not required to bear sole responsibility for retarded children, we are able to care for them without false heroism and overprotection and help them become responsible for their own lives. These consequences lead Hauerwas to conclude that an account of children as gifts offers a more sensitive and more practical approach to questions about having and caring for retarded children than does the notion of choosing our children favored by Fletcher.

IV. CARING FOR THE RETARDED

Hauerwas' aim is to find an explanatory context, what he calls a story, which can appropriately direct our attitudes and responsibilities toward our children, especially retarded children. I believe that Hauerwas does show that his story is able to acknowledge the value of retarded children and to suggest appropriate kinds of care better than Fletcher's account of having children as a matter of individual choice.[3] As May has stressed [11], the birth of a retarded child is a particularly strong blow to our hopes and expectations as parents. Fletcher's claim that we should strive to have only "high quality" children can only intensify this blow. It may leave us with a sense of failure or guilt (Could we have prevented this?) and no clear reason to accept and nurture the child. In this context, institutionalizing the child or allowing it to die might appear to be the least burdensome alternative. Or, if we choose to care for the child, our sense of complete responsibility for bringing it into the world may lead us to try to protect it from *every* failure or disappointment.

In contrast to this approach, Hauerwas strives to give us a way to understand and accept the birth of a retarded child. If we view children as gifts, we should recognize that they will not always be what we need or expect. Rather, children force us to learn new ways to live; in particular, Hauerwas claims, they teach us how to love without manipulating or controlling. Unlike nineteenth century views of the retarded as "holy innocents," eternal children who should be protected and kept content ([1], p. 26), Hauerwas stresses the need to help retarded children, like any other children, discover their own interests and goals, make their own mistakes, and in general, live independent lives. It is important to note that Hauerwas is not suggesting that we abandon the retarded to their own devices. Instead, I believe that he is claiming that his approach favors one kind of care, namely, helping the retarded to grow and develop their own skills in what courts have called the least restrictive environment, over more protective and custodial kinds of care.

Hauerwas' emphasis on helping the retarded to live independent lives is consonant with increasing evidence that most retarded persons can, given proper education and support, achieve higher levels of self-sufficiency than was formerly believed, and many can adapt to life in the community ([12], p. 138). This approach, however, seems least appropriate for the small percentage of the retarded whose developmental capability is so severely impaired as to make them unavoidably dependent on others for meeting their most basic needs. This group of severely and profoundly retarded persons will continue to require custodial, nursing, and frequent medical care.

Another suggestive feature of Hauerwas' account is his emphasis on the role of the community. By serving as a source of support and meaning for children apart from their parents, the community helps children to gain a measure of independence. Although Hauerwas has in mind a common religious tradition, his account might also be extended to the larger social community, in view, perhaps, of the community's commitment to a common belief in the dignity and rights of all its citizens. Implicit in Hauerwas's approach is the claim that the religious community shares with parents a responsibility for helping children to develop and to achieve an independent position within the community. Extended to the social community, this approach grounds a social responsibility to help provide those forms of support (e.g., education, job and housing opportunities) which will enable retarded citizens to lead independent, useful lives within the community. Despite Fletcher's pleas, this certainly seems to be a much more humane approach than Fletcher's own emphasis on the community as a kind of enforcer of genetic purity. Hauerwas' view of the retarded as gifts is not without problems, however. I will now turn my attention to these.

V. PREVENTION OF RETARDATION

We noted earlier that Fletcher's major concern is with the prevention of retardation. In fact, this concern is often linked with concern for providing better services for the mentally retarded. The President's Committee on Mental Retardation, for example, has formulated the major issues facing advocates of mentally retarded persons in the last quarter of this century in the following three questions:

(1) "Can mental retardation be significantly diminished as a human problem?"

(2) "Can the mentally retarded person be accepted as a citizen member of the community?"

(3) "Can humane services be so effectively delivered that the retarded person actualizes his full potential for human living?" ([12], p. 264).

Hauerwas' support for the objectives expressed in the latter two of these questions has, I hope, been made clear by the preceding discussion. Whether Hauerwas would also endorse the objective contained in the first question, however, is not so obvious. If the retarded are to be accepted as gifts from God, may we nevertheless attempt to prevent mental retardation? If so, may we use all the means suggested by Fletcher? (Among the means cited by the President's Committee on Mental Retardation are research into the causes of retardation and the avoidance or minimization of risks through public educational, social and health services – including genetic counseling and abortion.) Since he holds that we should wish no unnecessary suffering or pain on anyone ([8], p. 634), Hauerwas would likely favor those efforts (including research, education, social services and health care) which help us to conceive and bear healthy children. In contrast to Fletcher, however, Hauerwas objects to amniocentesis and selective abortion to prevent the birth of retarded children ([8], p. 632). His approach also raises doubts about methods such as genetic screening and counseling which would identify couples who risk conceiving retarded offspring, and thereby discourage them from having children. Both the abortion of affected fetuses and the decision not to conceive when there is risk of retardation become, in Hauerwas' view, an abrogation of the Christian duty to accept children, including retarded children, as gifts from God. This duty must be of fundamental importance for Hauerwas, since it apparently overrides concerns voiced by Fletcher about the possible suffering of the child, familial hardships, and social costs in caring for retarded children. These concerns are, for Hauerwas, presumably secondary in relation to the divine will that a child come into the world. Despite conflicts of values, therefore, procreative decisions are to remain largely in the hands of God, and efforts to prevent retardation must not interfere with God's procreative plans.

VI. RETARDATION AND DIVINE WILL

By making children gifts from God and reserving procreative decisions for divine, not human wills, Hauerwas' approach raises a broader theological question, namely, "Why does *God* create retarded people?" In fact, Hauerwas uses this very question to introduce his paper. He quotes it as part of a letter to the 'Wise Man's Corner' of the *St. Anthony Messenger*, a Catholic magazine. The letter reads:

How does one believe in God, Who is supposedly good, when there is so much unhappiness in the world? I have a mentally retarded sister who is in a state institution. On every visiting day it tears me apart to see such ugliness. I know I will never understand God's purpose in allowing these poor human beings to exist with the resulting heartbreak it causes their families every day of their lives. I do very much want to believe in God, but I guess my sister's existence has caused me to resent Him. How do I believe? ([8], p. 631).

Hauerwas maintains that this letter is theologically naive, since "it mistakenly assumes that God is to be held directly responsible for every unfortunate event that occurs in the world." He uses the letter to illustrate why we need a substantive answer to the question "Why should we have retarded children and welcome them into our lives?" Yet his answer to the question, namely, that we should accept retarded children as God's gifts, belies his initial claim that it is a mistake to hold God responsible, if not for every unfortunate event, then at least for the unfortunate events cited in the letter, namely, the suffering caused by retardation. Hauerwas' answer makes the writer's question even more urgent, since it acknowledges that the presence of the retarded is God's will, and makes human beings the duty-bound recipients of these divine gifts.

This problem is, of course, not new; rather, it is a version of the traditional theological problem of evil. Nevertheless, because the problem remains one of the most serious obstacles in the way of religious faith, and because Hauerwas' approach to the retarded raises the problem in an interesting way, I will pursue it, at least briefly. How might Hauerwas respond to the letter writer's plea for some explanation of God's intentions?

One strategy would be to reject the question, to affirm that God's purposes are indeed mysterious and human beings should not presume to understand them, but merely accept them. Although it may be most compatible with Hauerwas' position, this strategy is unlikely to help the questioner come to faith. It simply does not address the questioner's concern that the God whose goodness we assert also seems to be responsible for a condition that causes tremendous suffering and pain. How can a good God be responsible for this, or if God is responsible, in what sense can he be good?[4]

A different response would be that the suffering and pain which accompanies retardation, and perhaps even retardation itself, is not caused by God, but by human beings, through human negligence, self-indulgence, prejudice, insensitivity and other vices. According to this response, the suffering produced by retardation and perhaps existence of retardation itself could be prevented, but persists through human failures. This, incidentally, would probably be Fletcher's response. That is, he claims that since we have the technological capability to prevent handicaps, it is our responsibility to do so.

Not surprisingly, however, this second response fits poorly with Hauerwas' position. Although he does not explicitly say so, Hauerwas appears to hold that God does will *retarded* children, that is, retardation is not always the result of human failings. Indeed, some causes of retardation (e.g., undiscovered random mutations) are outside human control, and some measures to prevent retardation are unacceptable, given Hauerwas's assertion of a Christian duty to have children. Moreover, although much of the suffering which accompanies retardation is due to human insensitivities and could be ameliorated, at least some suffering (e.g., the frustration of having limited abilities) seems intrinsic to being retarded. Thus, if retarded children are created by God, He still appears to be responsible for at least some of their suffering.

Finally, a third response might acknowledge that God is responsible for retarded children and their suffering. The presence of the retarded, however, might be viewed as essential for the moral development of the communities in which they live. That is, the retarded might foster the virtues of benevolence, kindness, and tolerance. For example, in an address to the parents of retarded children, Hauerwas has claimed that "... without you and your children our communities would be less rich in the diversity of folk that we need to be good communities" ([9], p. 22).

Though this is perhaps the strongest of the above three responses, it too raises certain difficulties. If one holds that the retarded exist for the good of others, then the retarded have a primarily instrumental value.[5] In other words, God *uses* the retarded in order to produce better communities. This response leaves open the question whether the moral growth of the community is worth the cost of the involuntary suffering of some retarded children.[6] Moreover, our notion of divine justice appears shaken if God wills that some suffer so that others may benefit. In sum, the letter Hauerwas quotes raises a serious question for his position, a question which Hauerwas does not address and which has no simple solution.[7]

VII. CONCLUSION

In this paper, we have reviewed and evaluated two very different accounts of our responsibilities regarding persons with diminished natural abilities. Joseph

Fletcher advocates the unfettered use of new techniques to control the reproductive process and prevent the birth of diseased and retarded children. In contrast, Stanley Hauerwas emphasizes restraint, focusing on the divine role in creating children, including retarded children, and a corresponding human obligation to accept and nurture God's gift of children.

Each of these accounts, however, has certain drawbacks. Fletcher does not address the issue of what kind of care is owed to the retarded, and, as Hauerwas points out, his approach to childbearing may create a number of misconceptions about the nature of that care. Moreover, Fletcher's negative comments about retarded and handicapped individuals betray an inadequate appreciation of the value of their lives and may even contribute to prejudicial attitudes toward retarded persons. Hauerwas' position invites the question of why God should will that some individuals be retarded, and, as Fletcher would surely point out, it does significantly restrict the scope of human reproductive decision making.

I believe that a middle ground between these two positions, though probably unsatisfactory to both Fletcher and Hauerwas, would be most defensible. The greatest strength of Hauerwas' view, I think, is his insistence that God assumes a special responsibility for children. That is, children are not simply products at the disposal of their parents; rather, God seeks for them an independent existence within the larger community. Parents and communities, therefore, are called to help their children achieve their independence. This account, I suggest, guides our attitudes toward retarded persons more appropriately than Fletcher's account, which places God far in the background and asks persons to determine and to follow the loving, that is, the utility-maximizing course of action. Moreover, Hauerwas' emphasis on community support for the retarded affirms their value as persons in a way that Fletcher's emphasis on the community as enforcer of reproductive controls does not.

Hauerwas' position, however, connects God's special concern for children with divine responsibility for retardation and its attendant suffering. As the letter from the *St. Anthony Messenger* illustrates, ascribing this responsibility to God can become an obstacle to faith. This connection between divine concern and divine responsibility for retardation is not a necessary one; rather, it presupposes a particular view of the relationship between God and the world. Other views of God's relationship to the world would allow these two claims to be separated. That is, if complete control over worldly events is *not* attributed to God, then retardation can be viewed as caused by forces *in* the world, both human and non-human, and not by God. If retardation is viewed in this way, we can assert God's special concern for children without making God responsible for the presence of retardation. This would, of course, imply rejection of the traditional doctrine of absolute divine omnipotence. But other conceptions of the divine power are available; they have been developed in order to allow for genuine novelty within the world as well as for unavoidable tragedy and loss.[8]

If, as I have suggested, we acknowledge that God does not will that children be retarded, then I think it is also possible to take several steps in the direction advocated by Fletcher. That is, we are able to give human beings a greater opportunity to prevent retardation than Hauerwas allows. If retardation is a condition which neither God nor human beings would wish on anyone, then perhaps the Christian duty to have children should not extend to couples who discover that the risk is great that any children they might conceive would be retarded. Perhaps, also, the duty to accept children might be overriden when a couple discovers that their fetus's condition will cause it and them significant suffering. In these situations, I would suggest that more responsibility for making reproductive decisions can be granted to human beings without denying God's concern for retarded children or our duty to care for them appropriately. Though persons may sometimes legitimately choose to prevent the conception or birth of a seriously handicapped child, this would still be a far cry from the comprehensive, socially sanctioned, eugenic program proposed by Fletcher.

East Carolina University School of Medicine
Greenville, North Carolina

NOTES

[1] See, for example, ([13], pp. 26–33), [15].
[2] Fletcher takes a long step in this direction by advocating the use of donor sperm, ova, and embryos in order to choose 'one's children's biological quality' ([5], p. 161).
[3] Some of Fletcher's comments, however, suggest that he might simply deny that the lives of retarded children have positive value. Presuming happiness to be his utilitarian standard of value, however, such a denial seems false.
[4] I agree with Churchill's response that some mystery must be affirmed "without wallowing in mystery" ([2], p. 179). The question remains, however, whether the existence of the retarded is one of the things we must accept as mysterious.
[5] In his commentary, Churchill reminds us of God's instrumental use of Abraham, Isaac, Moses, Paul and Jesus ([2], p. 180). There is, however, a difference between these men and retarded persons. Abraham, Moses, Paul and Jesus freely respond to God's call to do his will. Isaac almost becomes an unwilling sacrifice, but is saved by God's intervention.

[6] Churchill takes issue with my use of the term 'involuntary' here ([2], p. 180). I grant that retarded people themselves cannot choose the suffering of retardation. This approach to the problem, however, would maintain that retardation *is chosen* by *God* in order to satisfy larger purposes. Thus, the retarded can claim that they had no say in God's choice of suffering for them.
[7] Churchill charges that I improperly impose the problem of evil, in its standard philosophical form, on Hauerwas's position ([2], p. 179). But the problem is, I think, clearly raised by the writer of the letter Hauerwas quotes. (Its significance for this issue is also recognized by May [11], p. 155.) Because Hauerwas does not address this problem himself, I consider several ways he might respond to the letter writer. These responses are sketched only very briefly; they do not delve into various specific approaches to the problem preferred by different theological traditions. Nevertheless, I hope that the responses are long enough to suggest some difficult and fundamental choices all face in providing a compelling solution to this problem.
[8] See, for example, ([7], pp. 134–138).

BIBLIOGRAPHY

[1] Allen, D. E. and Allen, V.: 1979, *Ethical Issues in Mental Retardation*, Abingdon Press, Nashville.
[2] Churchill, L.: 1982, 'Philosophical and Theological Perspectives on the Value of the Retarded: Responses to William F. May and John C. Moskop', in this volume, pp. 177–182.
[3] Darling, R. B.: 1979, *Families Against Society: A Study of Reactions to Children with Birth Defects*, Sage Publications, Beverly Hills, California.
[4] Fletcher, J.: 1966, *Situation Ethics*, Westminster Press, Philadelphia.
[5] Fletcher, J.: 1974, *The Ethics of Genetic Control*, Anchor Press/Doubleday, Garden City, New York.
[6] Fletcher, J.: 1979, *Humanhood: Essays in Biomedical Ethics*, Prometheus Books, Buffalo, New York.
[7] Hartshorne, C.: 1948, *The Divine Relativity*, Yale University Press, New Haven, Connecticut.
[8] Hauerwas, S.: 1976, 'Having and Learning How to Care for Retarded Children: Some Reflections', *Catholic Mind*, April 1976, 24–33, reprinted in S. J. Reiser, S. J. Dyck, and W. J. Curran (eds.), *Ethics in Medicine*, MIT Press, Cambridge, 1977, pp. 631–635.
[9] Hauerwas, S.: 1977, 'Community and Diversity: The Tyranny of Normality', *Newsletter of the National Apostolate for the Mentally Retarded*, Spring–Summer, 20–22.
[10] Lappé, M.: 1972, 'Moral Obligations and the Fallacies of "Genetic Control"', *Theological Studies* **33**, 411–427, reprinted in S. J. Reiser, S. J. Dyck, and W. J. Curran (eds.), *Ethics in Medicine*, MIT Press, Cambridge, 1977, pp. 356–363.
[11] May, W.: 1982, 'Parenting, Bonding, and Valuing the Retarded', in this volume, pp. 141–160.

[12] President's Committee on Mental Retardation: 1977, *Mental Retardation: Past and Present*, U.S. Dept. of Health, Education and Welfare, Washington, D.C.
[13] Rawls, J.: 1971, *A Theory of Justice*, Belknap/Harvard University Press, Cambridge, Massachusetts.
[14] Will, G. F.: 1980, 'The Case of Phillip Becker', *Newsweek*, April 14, 112.
[15] Williams, B.: 1973, 'A Critique of Utilitarianism', in J. J. C. Smart and B. Williams, *Utilitarianism: For and Against*, Cambridge University Press, Cambridge, England.

LARRY R. CHURCHILL

PHILOSOPHICAL AND THEOLOGICAL PERSPECTIVES ON THE VALUE OF THE RETARDED: RESPONSES TO WILLIAM F. MAY AND JOHN C. MOSKOP

I

Professor May's essay is designed to locate the valuing of the retarded in the context of parenting and bonding [4]. Concrete relationships, he asserts, are a necessary ingredient of our thinking. Whether these relationships constitute a sufficient context for decisions about the retarded is an important question, to be addressed later.

May correctly points out that the phrase 'natural abilities' commits us to what he calls a possessional view of persons which does not serve us well. He claims that such an abstract view of personhood leads to a minimalist set of responsibilities to others. May does not, however, indicate in detail which "disinterested judgments" about the retarded would tend to be minimalist in obligation. He might have said, as he has eloquently said in other places, that the resulting moral norm for the "less fortunate" is philanthropy, and that this sours into condescension and smacks of moral arrogance [3].

May speaks forcefully of both institutional and internal obstacles to bonding and his explication of the moments in the bonding process is quite useful. The notion of the child as a stranger I find especially provocative.

Several questions, however, remain and they are basic to issues May raises. Despite its elucidation of bonding, the thesis presented is essentially negative. May asserts that the full measure of the being and value of the retarded does not surface in a bondless consideration of their worth. This is an important thesis in its own right, but it leaves us somewhere short of a full affirmative position. The perceptions of those most deeply involved, May urges, must be a part of any decision process about care of the retarded. He eschews "disinterested judgment" and the perspective of anyone free of reciprocal relations with the retarded. The etymological play on the roots of the term 'value' makes this point effectively. But, like the rest of his essay, May's excursion into etymology portrays a scene rather than posing an argument. The portrait as a whole I find a powerful one, illumining dark corners of ourselves and our relationships. Yet it is still less than satisfying.

Recognizing the potential abuses of the bonded situation, May is chary of the power asymmetry between parents and children generally, and parents

L. Kopelman and J. C. Moskop (eds.), Ethics and Mental Retardation, 177–182.

and retarded children especially. This recognition seems to make him ambivalent about a forthright endorsement of the normative character of the parental perspective. So the question remains: What is the appropriate context for value judgments about the retarded child? We are told we cannot avoid the viewpoint of the bonded parent, but May stops short of making that viewpoint normative.

At the close of his essay May hints at, but does *not* elucidate, the perspective from which he believes the full being and worth of the retarded does emerge, viz., an encompassing God-relationship. Presumably parenting and bonding are not unrelated to the authorization and judgment of the more primordial God-relationship. Yet the latter is not explored but only suggested.

These limitations, however, do not diminish the force of May's essay, and they may suggest that he believes that enabling us to see what he shows is sufficient argument. Perhaps relational anthropologies cannot be argued for in the usual ways but can only be displayed. If this is the case, some methodological remarks would be in order. I hope Professor May will address this in the future.

In summary, Professor May's essay embodies dimensions of thinking about valuing the retarded that call for future study and elaboration. These are of both a methodological and a substantive sort.

May reminds us of the inherent prejudice in our own (or any) formulation of the question of worth of the retarded. In likening the difficulty of relating to the retarded 'thou' to the difficulty in relating to the sacred 'thou' we are provided a rich framework for understanding our perceptual biases. As we would frame (and have framed) a bourgeois God so are we tempted to frame a bourgeois regard for the retarded. Even the designation 'the retarded' so biases our perception. In this sense, our language and the values it bespeaks say more about us than about the retarded. The corrective is not to attempt to see from a neutral locus of perception (since none is humanly possible), but to attend more carefully to the sorts of relationships in which what can count as valuable is not so prejudicially formed.

To the extent that this can be argued for by being made visible to us, May is to be congratulated. To the extent that such making visible is precursory to more adequate policy formulations, May leaves himself, and all of us, with additional important work.

II

Professor Moskop's essay, by comparison, is directly theological and deals with the problem of evil [5]. Or more specifically, the subject matter he addresses is theological, as are the authors he considers, yet the mode of approach is decidedly philosophical. I mean by this that the problem of evil is a philosopher's problem. The problematic character of evil, in this vein, has to do with the inconsistency or contradictoriness of evil when coupled with an omniscient and omnipotent God of love. Power, foreknowledge, and beneficence are all necessary to crank up this problem, and all in pure degree and unlimited portions. The difficulties and the resolution Moskop presents are philosophical rather than theological, and to this extent they are not problems which those inside the circle of faith are likely to find convincing. I do not suggest some absolute chasm between philosophy and theology. I do not suggest that the problems raised are not real problems. They are genuine. I do argue that the mode of formulation of choices and the answers proposed will be different if one speaks, convictionally or confessionally, from within some Christian tradition (there are many and diverse traditions, not to be lumped together under a single label).

Let me develop my thesis by examining the three responses Moskop presents to the problem of evil. He finds flaws in all three responses. I believe all three responses would be affirmed from the confessional viewpoint, but not in the way Moskop suggests.

First, God's works *are* mysterious, but the inadequacy of human understanding of God is not as problematic as Moskop thinks. If understanding were complete, faith would be unnecessary; indeed, it would be reduced to magical superstition in Malinowski's sense or illusion in Freud's sense. But affirming mystery does not imply a naive piety or the anti-intellectualism Moskop suggests. One can affirm elements of mystery based in the transcendent nature of God without wallowing in mystery in a wholesale fashion.

Secondly, the response that suffering and evil are human rather than divine creations would be unsatisfactory only if it sets up a dualist view in which God is responsible for good and not evil. Such a response would, indeed, compromise the divine omnipotence. Moskop fails to mention a commonly held view which does not have this shortcoming, viz., the notion of a self-limiting God who allows freedom in man and indeterminacy in the natural order. Under such a scheme God is still responsible for evil, but presumably with a larger and beneficent purpose in mind.

And this takes us to the third response, viz., that the presence and suffering of the retarded is allowed by God in order to foster the development of certain prized human virtues, such as tolerance and humility. Moskop finds this objectionable because:

(1) it accords to the retarded a primarily instrumental value;
(2) moral growth may not be worth *involuntary* suffering; and,
(3) it shakes our notion of divine justice.

Let me take each of these objections in turn. I will argue that none of the objections are damaging because of Moskop's neglect of elements within traditional confessional perspectives. By failing to note how these problems would be posed differently from within a convictional belief system, Moskop has undermined the force of his objections. Moskop is, of course, free to pose the issues any way he sees fit for himself, but to the extent that he wants his formulations to be genuine responses to the believer he must take account of how the believer would pose the issue and what sort of answer would be satisfactory. The polemical context for Moskop's essay, it should be remembered, is Hauerwas' response to a troubled fellow Christian.

(1) The instrumental value of persons in God's larger scheme is a familiar theme in scripture. Consider Abraham, Isaac, Moses, Paul or Jesus, all of whom appeal to this mode of explanation. The purely instrumental value of personhood may be objectionable from the Kantian, Enlightenment view, but not to the providential schemes of the ancient Semites or first-century Christians. This is not to say that we should treat other persons as means rather than ends in themselves. Most religious teachings set standards more rigorous even than Kant's. Yet to affirm the intrinsic value of other persons does not invalidate the perception that God uses persons for larger purposes.

(2) Moral growth and suffering are notoriously difficult to measure. We cannot put them on a Kohlberg scale. From what advantage-point we might say "It wasn't worth it!" is unclear, and confessionally to assume such a sovereignty over judgments of worth would constitute moral hubris.

A second and related point is that suffering is not necessarily evil, though some suffering surely is. The suffering of the retarded and their families may or may not be evil. Moreover, the suffering of the retarded and their families is not involuntary in the sense of *not chosen*, but rather is a phenomenon which is *not chooseable*. With the possible exception of screening and amniocentesis, the suffering of retardation is neither chosen nor not chosen, neither voluntary nor involuntary. A better term is non-voluntary, designating an item over which "choice" is not a meaningful category at all. That there are aspects of life over which choice plays no part is an unseasonable thought in a culture which has assumed that the discrete acts of choosing are the sum total of morality – a paradigm for the moral life – and that deciding and choosing all-but-exhaust ethics. The assumption that voluntary suffering is morally less objectionable indicates the centrality of choice in our contemporary ideas about the moral life. While this is an accurate portrayal of contemporary moral philosophy, choosing is only one dimension in most religious moral systems, which emphasize the "being" side of morality as expressed in virtues and character.

(3) Lastly, divine justice I understand as an eschatological category, not a self-evident or purely temporal notion. We succumb to what Mircea Eliade terms the idolatry of history if we locate God's justice in history and make it accountable to our own sense of justice. I take it that a part of the force of any notion of divine justice is as a corrective for the inadequacy of our more ordinary ideas of justice.

I suggest that from the confessional perspective the question is not "Have the retarded been wronged by having been created retarded?", but "Have the retarded been wronged by having been created?" The choice is not between perfection and imperfection, or normality and abnormality, but between being and non-being. One of the most fundamental of religious experiences is *awe*, evoked in our recognition that people exist at all. Or if one prefers a philosophical confessional, "It is not *how* things are in the world that is mystical, but *that* it exists" ([6], p. 149). The major theological thrust of liturgy and sacrament is that existence itself is an undeserved favor.[1]

The difficulty Moskop's essay embodies is not a surface misinterpretation, but a profound mismatch of language to experience. In its generic form it is the taking of the experiences of others and couching them in one's own idiom and categories of thought and treating those experiences – not as authentic in their own right – but as examples or specimens of a standard problem. In its specific form it is the taking of the experiences of the retarded and their parents and treating them as an example of the problem of evil, in its standard philosophical form. The difficulty is that those elements which do not fit the stock formulation are ignored or overlooked. Much can be lost in the translation and the concrete experiences of actual people can easily become distorted as "the problem" is solved.

William James called this difficulty "vicious intellectualism,"[2] the admission into legitimacy of only those aspects of the issue which conform to one's preconceived notions. This is all the more powerful if, as is the case here, it is done with the best of intentions and unselfconsciously. This difficulty is, of course, inherent in all intellectual work and in any use of categories of understanding. The effort to make conceptual formulations as commensurate as possible with actual experiences is a basic requirement of all academic studies.

What sort of problem is the problem of evil? For whom is it a problem? To the philosopher and theologian alike, to be sure, but in different ways. I have suggested that the problems Moskop raises would be problems for Hauerwas as well, but they would not be problematic in the way Moskop suggests. This divergence may or may not apply to Joseph Fletcher, whose leanings are more congenial to act utilitarianism than to the sort of theological reflection which is deeply informed by religious traditions. This is not to say that Moskop's own formulations and their resolution are not legitimate in their own right, as issues for him. I look forward to the further development of his own thought. We can only hope that theologians will not mistake his enterprise for confessional theology and thereby distort the history and context in which his thinking belongs and outside of which it risks losing coherence.

School of Medicine
The University of North Carolina at Chapel Hill
Chapel Hill, North Carolina

NOTES

[1] For a first-rate philosophical explanation of this point, see [1], pp. 324–326.
[2] James's formulation is to be found in [2], p. 60. "The treating of a name as excluding from the fact named what the name's definition fails positively to include . . .".

BIBLIOGRAPHY

[1] Adams, R. M.: 1972, 'Must God Create the Best?' *Philosophical Review* 81, 317–332, reprinted in this volume, pp. 127–140.
[2] James, W.: 1909, *A Pluralistic Universe*, Longman, Green and Company, New York.
[3] May, W.: 1977, 'Code and Covenant or Philanthropy and Contract', in S. J. Reiser *et al.* (eds.), *Ethics in Medicine*, MIT Press, Cambridge, Massachusetts, pp. 65–76.
[4] May, W.: 1984, 'Parenting, Bonding and Valuing the Retarded,' in this volume, pp. 141–160.
[5] Moskop, J.: 1984, 'Responsibility for the Retarded: Two Theological Views,' in this volume, pp. 161–175.
[6] Wittgenstein, L.: 1961, *Tractatus Logico-Philosophicus*, D. F. Pears and B. F. McGuinness (transl.), Humanities Press, New York.

SECTION IV

LAW AND PUBLIC POLICY

MICHAEL KINDRED

THE LEGAL RIGHTS OF MENTALLY RETARDED PERSONS IN TWENTIETH CENTURY AMERICA

The United States Supreme Court in 1972 freed a mentally retarded person from lifetime court-ordered confinement. In *Jackson v. Indiana* [53] the Court accomplished this result by articulating the modest proposition that "due process requires that the nature and duration of commitment bear some reasonable relation to the purpose for which the individual is committed" ([53], p. 738). Jackson, who was a mentally retarded deaf-mute, had been charged with two thefts of small amounts of property. Because of his handicaps, he was found incompetent to stand trial. Without any determination of guilt or innocence he was ordered by the Indiana court to be confined by the Department of Mental Health until "sane" ([53], pp. 717–719). The Supreme Court held that the nature and duration of Jackson's confinement was so devoid of a legitimating purpose that the confinement was an unconstitutional denial of due process ([53], pp. 736–739).

In addition to this "due process" basis for its decision, the Court had an "equal protection" basis. It compared the criteria by which *Jackson* was committed with Indiana's civil commitment criteria and civil commitment release criteria. It found differences in the commitment and release criteria to be significant and to be lacking in rationality ([53], pp. 727–729). Using an approach developed in several earlier cases involving mentally ill persons [44, 51], the Court held that these irrational differences amount to a denial of equal protection for a person in *Jackson's* position ([53], p. 730). On these two bases, "due process" and "equal protection," *Jackson* invalidated a long-standing process ([6], pp. 408–423) through which states confined many mentally disabled persons ([11], p. 66) for life in terrible prison-asylums ([17], p. 843; [28], p. 456) for nothing more than having been charged with a crime and found "incompetent to stand trial" ([6], pp. 408–423).

Jackson v. Indiana may stand for some time as the high-water mark in the ebb and flow of American legal treatment of the mentally retarded, and so is worthy of note. It is also noteworthy because of the thoughtless irrationality and cruelty exposed in a process that was utilized in many states. Finally, it is significant that the bases for decision are so general. While broad concepts permitted the Court to strike down the absurd criminal

L. Kopelman and J. C. Moskop (eds.), Ethics and Mental Retardation, 185–208.

incompetence confinement system as it operated to impose indefinite confinement, the case provides little guidance for future legislative, administrative, or judicial development. This generality is typical of the few Supreme Court pronouncements in favor of the rights of handicapped persons. No Supreme Court opinion goes further than *Jackson* in its protection of handicapped persons, and, as will be argued in Part III of this chapter, a number of the Court's more recent decisions are restrictive of the legal rights of handicapped persons.

This chapter focuses on the legal treatment of mentally retarded persons by the American legal system during this century, although a few of the cases discussed deal with handicaps other than mental retardation. Three significantly different approaches can be discerned. The period from the beginning of this century until well into the 1960s is characterized by legal neglect of the mentally retarded. Then, during the 1970s a strong contrast is provided by a "civil rights reform" period. Finally, it is noted that the United States Supreme Court has stayed fairly aloof from much of the civil rights reform litigation. Through a review of some very recent decisions, including several decided during the final writing of this chapter, the question of the Supreme Court's future direction is posed.

The chapter suggests an oscillation in the relationship between American society and its mentally retarded members, perhaps reflecting a tension between rejection of and identification with the limitations implicit in the concept of mental retardation ([36], p. 27). Such an oscillation now in attitudes and policies towards the mentally retarded would not constitute the first swing of the pendulum. The nineteenth century also witnessed a full swing. The early and middle portions of the nineteenth century brought optimism and commitment [21, 3]. The first residential institutions for the mentally retarded in America were founded during this period as true boarding schools, in which mentally retarded persons were to learn skills that would permit them to return to the community in productive roles [3, 38, 21]. Well before the turn of the century, however, this optimism about mental retardation and about mental retardation institutions had shifted to a pervasive pessimism [3, 38, 21], ([42], p. 29). Public policy came to be characterized by programs of institutional segregation and eugenic sterilization [3]. Institutions became regarded as purely custodial [42]. The number of persons in institutions for the mentally retarded grew from 4000 in 1890 ([42], p. 410) to approximately 200,000 in 1970 ([3], p. 18). The number of persons so institutionalized per 100,000 Americans went from ten to a hundred in this period ([3], fig. 1–3, p. 19).

I. THE PERIOD OF NEGLECT

The bulk of this chapter will focus on aggressive litigative, legislative, and administrative initiatives on behalf of mentally retarded citizens during the 1970s. Nevertheless, it is essential to put this brief period of ferment in context by focusing first on the half century that preceded it. That period is characterized by an eerie silence in the legal system. Not much was said by the courts, the legislatures, or legal scholars. What was said may help explain the absence of other voices.

The Supreme Court of the United States addressed the question of the legal rights of the mentally retarded only once during this period, and then in a voice that could quiet protest for decades to come. In 1927 the Court decided *Buck v. Bell* [47], upholding a Virginia statute that permitted the involuntary sterilization of inmates of named Virginia institutions who were "afflicted with hereditary forms of insanity that are recurrent, idiocy, imbecility, feeble-mindedness or epilepsy" upon the recommendation of the institutional superintendent and concurrence of the institution's board of directors [75].

The statute was challenged on two principal grounds: (1) that "substantive due process" was denied in that the state lacked the power to invade any citizen's bodily and procreative integrity; and (2) that "equal protection" was denied in that the statute applied only to institutionalized persons and did not cover similarly handicapped persons living outside the institution [8].

The due process challenge did not attack the procedures under the statute ([47], p. 207), which required an administrative hearing and permitted an appeal to a judicial forum [75]. Mr. Justice Holmes, writing for the Court, nevertheless described those procedures at length, saying they provided for very careful consideration of the patient's rights ([47], pp. 206–207). When Mr. Justice Holmes finally turned his attention to the substantive challenge to the state's power to act, he seemed impatient, almost dismissive. He said:

> We have seen more than once that the public welfare may call upon the best citizens for their lives. It would be strange if it could not call upon *those who already sap the strength of the State* for these lesser sacrifices, often not felt to be such by those concerned, in order to prevent our being *swamped with incompetence*. It is better for all the world, if instead of waiting to execute *degenerate offspring* for crime, or to *let them starve for their imbecility*, society can prevent those who are *manifestly unfit* from continuing their kind. The principle that sustains compulsory vaccination is broad enough to cover cutting the Fallopian tubes. . . . *Three generations of imbeciles are enough* ([47], p. 207, *emphasis supplied*).

The passage is permeated by factual and value assumptions. Mentally retarded citizens are apparently not "the best". Their "incompetence" risks swamping "us." They will have "degenerate" offspring. That the Court is irritated at being troubled by such a trivial matter is suggested by its analogy to vaccination. In fairness to Mr. Justice Holmes, one should note that Ms. Buck's attorney provided little argumentation to support his substantive due process position; he simply asserted that there is "an inherent right of mankind to go through life without mutilation of organs of generations (sic)" ([8], p. 501).

Even more telling than the Court's rejection of the due process argument was the treatment accorded the equal protection challenge, which focuses upon the categorization of persons under the statute. Here the statute singled out institutionalized mentally retarded persons for involuntary sterilization. The Court began by characterizing equal protection as "the usual last resort of constitutional arguments" ([47], p. 208). It then focused on the distinction between mentally retarded persons in institutions and those outside and cynically responded that any difficulty caused by statutory exclusion of the latter would be solved partially by the sterilizations accomplished under the statute. "So far as the operations enable those who otherwise must be kept confined to be returned to the world, and thus open the asylum to others, the equality aimed at [*i.e.* equality of opportunity for involuntary sterilization] will be more nearly reached" ([47], p. 208). Perhaps most revealing of all is that appellant's counsel himself did not challenge the classification drawn between those who are mentally retarded and those who are not [8]. He, too, focused on the discrimination among the handicapped against those in institutions ([8], p. 502). Indeed, counsel argued that sterilization was not required by the state's purpose because the goal of precluding conception was being accomplished by institutional segregation ([8], p. 508).

There were few if any Supreme Court opinions even mentioning mental retardation in the succeeding four and half decades. In 1972 the Supreme Court was to note its surprise that there had been so few challenges to the states's power to restrict the mentally disabled prior to that time ([53], p. 737). One may ask how surprising that dearth of challenges is when one sees how readily the Supreme Court dismissed the challenge in *Buck v. Bell*. In any case, one need not expect challenges if courts, families, and counsel alike share the assumption that restriction of mentally retarded persons is to be expected. A significant change in perspective would be required before courts would speak of the civil rights of mentally retarded citizens.

Buck v. Bell stands out as a dismissive judicial pronouncement on the place of mentally retarded persons in American society. Its importance is accentuated by the absence of offsetting judicial pronouncements. It is consistent with, and reinforced by, the approach taken in legislative acts during the period. At the federal level, Congress gave almost no attention to the mentally retarded until the late 1950s ([33], p. 43) and little until the 1970s. State laws dealt with the mentally retarded summarily, often carelessly, and generally with the effect of denying rights assumed by most citizens to be of fundmental importance.

Jackson v. Indiana [53], described earlier, documented the tragic results of such state legislative thoughtlessness. A totally irrational structure was allowed to develop on the basis of legislation that said simply "a person insane at the time of trial can be confined until his sanity is restored" [53, 6].

Such disregard was pervasive. As public education became well established, the mentally retarded were peremptorily excluded from it ([20], p. 252). The matter was often disposed of by a simple state statute authorizing the local board of education to exclude from school any child it found unable to profit from a public eduction ([10], pp. 63–68).

In the realm of guardianship, to take another example, the law often simply provided that upon a finding of 'incompetence' a guardian could be appointed. The guardian would substitute for the 'incompetent' in all legal decisions ([2], p. 34). In many states today guardianship statutes still depend on a global finding that a person is 'incompetent', with a consequent exclusion from any legal self-determining action. While legislation has been enacted in some states calling for more subtle judgments of capacity and limitations upon capacity ([10], pp. 573–574), it is questionable whether such legislation has been effectively implemented ([1], p. 19).

Although fairly elaborate legislative provisions were developed in the nineteenth century to govern the institutionalization of the mentally ill ([6], pp. 134–132), institutionalization of the mentally retarded was often dealt with much more summarily. In Ohio, for example, it was governed until recently by two very simple provisions. The first stated:

> The authority to . . . hospitalize . . . a person alleged to be . . . mentally retarded . . . shall be the same as is provided for the mentally ill and insane insofar as may be applicable to the mentally retarded. . . . The procedures set forth . . . with respect to hospitalization [of the mentally ill] shall also apply to and be followed in cases pertaining to mental retarded patients insofar as the same may be applicable ([74], § 5125.25 (repealed 1975)).

In case that did not "dispose of the problem," the legislature further provided for the eventuality that institutions were too full to accept additional residents. It said:

> The probate judge shall then take such action . . . as he deems necessary and advisable to provide for the detention, supervision, care, and maintenance of said mentally retarded person, at the expense of the county, until such time as he may be received in a hospital for the mentally retarded ([74], § 5125.30 (repealed 1975)).

It will be noted that the probate judge is thus given almost unbridled discretionary control over mentally retarded persons. The discretion is limited by the state statute in only one way: any costs must be born by the county, not the state. In essence, the legislature said, "This population is not worth the thought it would require to legislate appropriately for it."

II. THE DECADE OF CIVIL RIGHTS REFORM

The contrast between the dismissive first half of the twentieth century and the reformist – and regulatory – seventies is striking ([32], p. 3). The dismissive earlier period has been insisted upon in part because of some uncertainty as to whether it is in fact over. Nevertheless, the era of civil rights reform transformed mentally retarded persons – at least for a time – from objects of the law to citizens with rights [22]. A caution, however, is still in order. In spite of some slight Supreme Court participation in the reforms of the seventies [53, 57, 43], by far the greatest part of the movement has occurred in the federal trial courts, the district courts. If one looks only at Supreme Court cases, it is much less clear that a decade of civil rights reform has occurred. The Supreme Court has been reluctant to subscribe to the reformist positions of the trial courts. This decade of so-called 'civil rights reform' may appear in deeper retrospect as a mere aberration, a lower court revolt, once the Supreme Court has turned its full attention to the matter.

The sweep of reform in the federal district courts can and will be illustrated by several major cases, each of which attacked a major facet of society's treatment of the mentally disabled. These cases are *Pennsylvania Association for Retarded Children (P.A.R.C.) v. Commonwealth of Pennsylvania* [60], *Lessard v. Schmidt* [54], *Wyatt v. Stickney* (later *Aderholt*) [67, 68, 69], and *Halderman v. Pennhurst State School and Hospital* [49, 50, 59]. The *P.A.R.C.* case attacked the exclusion of mentally retarded children from the public school system. *Lessard* attacked loose civil commitment procedures. The *Wyatt* and *Pennhurst* cases attacked the warehousing of mentally retarded persons in large custodial institutions. In each case, the abuse attacked was obscene. Once scrutinized in court, a remedy was ordered.

These cases will receive further examination, but first one can well ask why the decade of the seventies was the moment for American society, through its judicial system, to scrutinize these long-standing abuses. The answer to this question of timing must be largely speculative, since it is a question of social causation. Nevertheless, it is possible to note occurrences that may have affected this development.

One can wonder whether an early role belongs to German social planners of the nineteen thirties. The eugenic movement's premises were that it was scientifically possible and ethically desirable to identify some human characteristics as inferior and to eliminate those traits through state action in order to 'improve' the 'human race' [25, 19]. These premises provided the logical basis for sterilization statutes in a number of American states [16, 9], for the decision of the United States Supreme Court upholding such statutes in *Buck v. Bell* [47], and in part for the concentration of mentally retarded persons in large sexually segregated warehousing institutions ([42], pp. 38–53). The movement's targets in the United States were not limited to the mentally retarded. A "Model Eugenic Sterilization Act" published under the aegis of the Chicago Municipal Court in 1922 called for the sterilization of "potential parents of socially inadequate offspring" ([25], pp. 446–449). It singled out immigrants as a special problem and proposed giving the federal government jurisdiction to sterilize immigrants who were potential parents of socially inadequate offspring ([25], pp. 451–452).

When Nazi Germany enacted a eugenic sterilization law and set up sterilization courts in 1934, the magazine *Scientific American* published a series of four articles debating the wisdom of the legislation [24, 19, 40, 13]. American writers argued both pro [19] and con [13]. A German spokesman began his defense of the law by claiming that Germany was simply joining with thirty American states in this endeavor [40]. It may at least be that the Nazi demonstration of the dangers inherent in any effort to draw lines between humans with rights and those without them, or between more and less worthy types, classes, or races was a remote source of the civil rights decade for the mentally disabled ([37], Intro. p. xxii), [23, 9], ([14], p. 1456). Perhaps American revulsion at the consequences of such policies aided in a redefinition of American values that made it clear that our survival as a culture depends upon integrative ([18], p. 174) rather than segregative solutions and upon the values of racial diversity and acceptance rather than racial purification.

One can wonder if this is also an attitudinal issue on which there is periodic oscillation. The author of this chapter was uncomfortable with the extent to which some speakers at the conference from which this book is derived found the relevant focus to be on which individuals conceived by human parents are to be excluded from the rights-holding polity. One author has stated, in what seems like a monumental understatement, that the Nazi example makes it imperative that in any new eugenics program "[t] he choice of who is to select the traits to be diminished and those propagated must be made so as to insure fairness to all" ([41], p. 201).

A second possible cause for the change of approach is political. The parents of mentally retarded children formed organizations after the Second World War ([35], p. 53). As these parent organizations progressed, they developed mutual self-help programs that demonstrated that in fact there were alternatives to locking away and sterilizing the mentally retarded. These mutual aid associations also emboldened parents to speak more openly about the existence of mental retardation and the needs of their mentally retarded family members ([15], pp. 7–8; [37], Intro. p. xxiii).

A third event was the election of John F. Kennedy as President of the United States. President Kennedy's willingness to use his office to focus attention on the plight and prospects of the mentally retarded resulted in the appointment of a President's Panel on Mental Retardation, which published a report [34], and in a substantial increase in federal support for community-based services ([12], p. 3).

A fourth event is the civil rights movement for Black Americans. The denomination in this chapter of the period under discussion as the civil rights reform decade reflects the author's judgment that the civil rights movement was a key factor in shaping the development of initiatives that were to effect the mentally retarded ([18], note 5 and accompanying text, pp. 174–175). There are several aspects of the civil rights movement that seem to be important. Conceptualization in terms of "minority group" status and "discriminatory" action were critical to the civil rights movement for Blacks. In order for the civil rights movement on behalf of Black Americans to move ahead there had to be acceptance of the notion that the United States has not, from a political point of view, behaved as one grand continuum of individuals, but that on the contrary there are instances in which a majority has acted in ways that are highly prejudicial to the interests of a discrete minority. This provided a conceptual model for analysis of the plight of the retarded – as another minority group, also badly disserved by the actions of the majority. Another aspect of the Black American civil rights movement that was of importance to later civil rights efforts on behalf of the mentally retarded was increased sophistication in the use of governmental machinery to address social issues. Blacks discovered that the Constitution and courts, and particularly the federal courts, could be utilized successfully to expose discriminatory mistreatment and to remedy the continuing harm. There also developed a realization that federal court action had to be part of a broader strategy to effect change, a strategy involving federal legislative, regulatory, and executive activity as well as local and state political and governmental follow-up efforts. This sophistication in the workings of the American legal and political system was utilized in the 1970s to win major improvements in the legal treatment of mentally retarded persons.

Whatever the reasons, the 1970s saw a great expansion of litigation, legislation, and administrative action on behalf of mentally retarded (and other handicapped) persons. Reflective of this expansion is the birth of several specialized publications to report on current developments. The President's Committee on Mental Retardation commissioned a publication that appeared every few months during the 1970s entitled *Mental Retardation and the Law: A Report on the Status of Current Court Cases*. Congress funded a National Center on Law and the Handicapped, which published a monthly magazine called *Amicus*. The American Bar Association established a Commission on the Mentally Disabled, which in turn has published since 1976 a *Mental Disability Law Reporter*. Perhaps indicative of more recent trends, only the last of these three publications appears to have survived into the early 1980s. There also have been several symposia in law journals, and a substantial increase in the volume of scholarly attention to the topic.

With these general introductory comments, it is now possible to consider several areas in which the 1970s brought substantial, if temporary, change in the legal status of mentally retarded Americans.

The first of the civil rights reform era cases was *Pennsylvania Association for Retarded Children (P.A.R.C.) v. Pennsylvania* [60]. This suit challenged the exclusion of mentally retarded children from the public school system in Pennsylvania ([26], p. 25). Several different legal techniques were alleged to have been used by Pennsylvania school districts to exclude mentally retarded children from the public school system. Among these were an exception to the compulsory school attendance law for children found "unable to profit from further public school attendance" ([60], p. 1262) and a provision for the certification of children as "uneducable and untrainable in the public schools" ([60], p. 1264).

On August 12, 1971, the Federal District Court for the Eastern District of Pennsylvania heard testimony from four of plaintiff's expert witnesses. Following this presentation of testimony, the state attorney general entered into negotiations with plaintiff's attorneys ([26], pp. 29–30). The two sides reached a "consent agreement" on the issues raised and submitted this agreement to the court for its approval [60]. The court approved inclusion in the consent agreement of the following language:

> Expert testimony in this action indicates that all mentally retarded persons are capable of benefiting from a program of education and training; that the greatest number of retarded persons, given such education and training, are capable of achieving some degree of self-care; that the earlier such education and training begins, the more thoroughly and the more efficiently a mentally retarded person will benefit from it; and, whether begun early or not, that a mentally retarded person can benefit at any point in his life and development from a program of education and training ([60], p. 1259).

The consent agreement approved by the court required the State of Pennsylvania to provide a free and appropriate public education to all mentally retarded children between the ages of 6 and 21 ([60], p. 1259). The Attorney General agreed to issue interpretations of the compulsory attendance and certification laws that would reflect the fact that "all children are capable of benefiting from a program of education and training" ([60], p. 1265).

It is evident that a substantial reorientation is required to move from a system that rejects children as "uneducable" to a system that treats all mentally retarded children as susceptible to education. The reorientation involves either a different factual assumption about mentally retarded persons or a different conceptualization of education, or both.

The *P.A.R.C.* case was followed fairly quickly by a right to education decision that was not resolved by a consent decree, *Mills v. Board of Education of the District of Columbia* [55], ([20], pp. 257–262). Following these two decisions, right to education suits were filed in many school districts across the United States ([20], p. 262). In addition, the litigative breakthroughs of *P.A.R.C.* and *Mills* were consolidated quickly by federal and state legislative and administrative action.

On the basis of the *P.A.R.C.* and *Mills* decisions ([29], p. 113; [30], p. 1105), Congress passed the Education for All Handicapped Children Act of 1975 [72], which requires any state accepting federal education funds to provide a free and appropriate public education for all handicapped children ([30], p. 1105; [29], p. 123). In order to avoid litigative problems ([29],

p. 119) and to comply with the requirements for continued federal funding, many states revised their education statutes to assure the right to education for handicapped children as a matter of state law. Thus, by late in the 1970s the right to an education for mentally retarded children was generally accepted. The forum had shifted back from the federal to the local level and the issues were those of implementation rather than of basic entitlement [30].

The other major area of reform to be examined here is that of institutional segregation. By the 1960s, almost 200,000 mentally retarded individuals were confined in large institutions ([3], p. 20). Criteria and procedures for confinement were both loose, and institutional conditions were atrocious. *Jackson v. Indiana*, discussed in the introduction to this chapter, attacked some of the loose and irrational criteria for confinement. *Lessard v. Schmidt* called for wholesale revision of the procedures by which persons were committed to institutions for the mentally ill [54]. The *Lessard* case applied procedural notions of notice, hearing, burden of proof, and right to counsel to the civil commitment system.

Beyond the issues of criteria and procedures for placement in an institution, an even more fundamental question has received much judicial attention but only very limited Supreme Court attention. This is the issue of right to treatment or habilitation.

The first judicial declaration of a right to habilitation for institutionalized mentally retarded persons came in *Wyatt v. Stickney* (later renamed *Wyatt v. Aderholt*) [68]. Although this decision was preceded by two more limited decisions involving the right to treatment of individual mentally ill persons [61, 56] and by scholarly writing discussing whether a "right to treatment" for the mentally ill ought to be recognized [5, 27, 39], *Wyatt v. Stickney* (*Aderholt*) was still a landmark decision. It addressed fundamental questions that had been raised only a decade earlier [5]. Plaintiffs utilized expert witnesses to establish that the conditions in Alabama's public institutions for the mentally retarded were appalling ([68], pp. 391–394). Upon determining this fact, the Court decreed that:

> Because the only constitutional justification for civilly committing a mental retardate is habilitation, it follows ineluctably that once committed such a person is possessed of an inviolable right to habilitation ([68], p. 390).

This position was affirmed by the Fifth Circuit Court of Appeals [69].

The *Wyatt* case, having based the right to habilitation for the mentally retarded on the due process clause of the federal constitution, ordered extensive changes in the operation and staffing of the institution ([68], pp. 395–407).

Wyatt v. Stickney (*Aderholt*) was followed by numerous law suits attacking conditions in state institutions for the mentally retarded across the country. While a fair number of these suits have been settled through consent decrees, courts that have addressed the issue have generally concurred with the *Wyatt* decision that federal constitutional rights are violated by simply warehousing mentally retarded persons in custodial institutions. In the meantime, Congress passed legislation, entitled the Developmentally Disabled Assistance and Bill of Rights Act, in which it stated that it "found" that there is "a right to habilitation for the mentally retarded" ([71], § 6010).

One of the most recent right to habilitation law suits, *Halderman v. Pennhurst State School and Hospital* (*Pennhurst*) [49, 50, 59], spans and will provide a transition between this chapter's consideration of the civil rights reform decade and the present. Building on the *Wyatt* precedent and the subsequent federal legislation, the *Pennhurst* case challenged the legality of conditions at Pennhurst State School. As the suit evolved, it brought into question the very existence of Pennhurst, and by implication, of places like it.

The case involved a trial of thirty-two days ([49], p. 1298), in which the district judge became convinced that Pennhurst State School for the Mentally Retarded, a not atypical institution for the retarded and one that had been a fundamental part of the Pennsylvania scene for seventy years ([49], p. 1302), must be phased out under federal court order ([49], p. 1325). This is very drastic action for a federal court to take, and could not have been accomplished without the overwhelming factual demonstration that came from the long trial in the case. Plaintiffs' attorneys used a number of different kinds of evidence. Experts testified that the conditions at Pennhurst were deplorable and that it was not an habilitative setting for anyone ([49], pp. 1304–1308); experts also testified that community alternatives were feasible for even the most severely retarded persons and that they would be better served in these more integrative environments ([49], pp. 1311–1312). Individual cases were documented to demonstrate the broad propositions in personal terms ([49], pp. 1309–1310). A sampling of patient files was used to establish the generality of the lack of care and deterioration of the population during confinement in Pennhurst ([49], p. 1308, note 40).

The District Court judge made the following findings of fact:

> The average resident age at Pennhurst is 36, and the average stay at the institution is 21 years. Forty-three percent of the residents have had no family contact within the last three years. Seventy-four percent of the residents are severely to profoundly retarded. The average resident has had one psychological evaluation every three years and one vocational adjustment service report every 10 years. Those residents who have had more

than one Vineland examination (measuring social quotient) during their residency at the institution, have, on the basis of this test, shown a decline rather than an increase in social skills while at Pennhurst (they declined an average of 7.542 points during their residence at Pennhurst, a loss of 0.596 points per year.)

At its best, Pennhurst is typical of large residential state institutions for the retarded. These institutions are the most isolated and restrictive settings in which to treat the retarded. Pennhurst is almost totally impersonal. Its residents have no privacy – they sleep in large, overcrowded wards, spend their waking hours together in large day rooms and eat in a large group setting. They must conform to the schedule of the institution which allows for no individual flexibility. Thus, for example, all residents on Unit 7 go to bed between 8:00 and 8:30 p.m., are awakened and taken to the toilet at 12:00–12:30 a.m. and return to sleep until 5:30 a.m. when they are awakened for the day, which begins with being toileted and then having to wait for a 7:00 a.m. breakfast ([49], pp. 1302–03, footnotes and references to transcript omitted).

The District Court was convinced that a major legal response was necessary. It held that confinement by the state of mentally retarded persons in an inadequately habilitative and unnecessarily restrictive setting violates those persons' rights ([49], pp. 1313–1324). It held further that the evidence had established that Pennhurst could not serve as the locus of such habilitation and that its residents must be moved to less restrictive environments ([49], p. 1325). The Court appointed a master to oversee implementation of this broad directive ([49], p. 1326). The District Court held that plaintiffs' claim could be equally well based upon constitutional concepts of due process, freedom from cruel and unusual punishment, and equal protection, or on Section 504 of the Rehabilitation Act of 1973, or upon a Pennsylvania state statute ([49], pp. 1314–1324).

The State of Pennsylvania appealed this decision. The Court of Appeals for the Third Circuit deliberated *en banc* on the case. It chose not to consider the constitutional arguments ([50], p. 94), but rather based its decision on the federal Developmentally Disabled Assistance and Bill of Rights Act [71], ([50], p. 97) and on Pennsylvania state legislation ([50], p. 103). The Bill of Rights provision of the Developmentally Disabled Assistance and Bill of Rights Act states:

Congress makes the following findings respecting the rights of persons with developmental disabilities:

(1) Persons with developmental disabilities have a right to appropriate treatment, services, and habilitation for such disabilities.

(2) The treatment, services, and habilitation for a person with developmental disabilities should be designed to maximize the developmental potential of the person and should be provided in the setting that is least restrictive of the person's personal liberty .... ([71], § 6010).

The Court of Appeals also held that the District Court had gone too far in concluding that no one could be habilitated at Pennhurst. While the Court of Appeals agreed that the evidence was sufficient to create a presumption against continued placement of any individual in Pennhurst, it held that there would have to be individual hearings at which there would be an opportunity to demonstrate that a particular resident would be best served by remaining at Pennhurst ([50], pp. 113–114). The Court seemed particularly concerned about aged residents, for whom a move away from familiar surroundings might be traumatic ([50], p. 114).

In this posture, the case was appealed to the Supreme Court of the United States, whose treatment of this and other matters is the topic of Part III.

III. THE UNITED STATES SUPREME COURT – BACKLASH?

At the outset of this chapter, it was suggested that the United States Supreme Court has remained aloof from the development of legal rights for the mentally retarded and that *Jackson v. Indiana* might stand for some time as the the high-water mark in this development. Indeed, the cases the Court has considered that involve the mentally retarded are so few that this section will continue to consider some Supreme Court cases involving handicapped persons whose handicap is other than mental retardation. The hallmark of the Court's involvement has been caution to a point that could be characterized as negative. This is particularly true in several cases where the Court has given very restrictive interpretations of federal statutes.

As noted in the introduction to this chapter, one characteristic of the Court's approach in those cases where it has sustained the position of the handicapped individual has been extreme generality. The declaration in *Jackson v. Indiana* that "the nature and duration of commitment [must] bear some reasonable relation to the purpose for which the individual is committed" ([53], p. 738) was very broad. This is typical of the Court's favorable pronouncements.

The Supreme Court has been very cautious in its consideration of the rights of mentally retarded or mentally ill persons within state civil institutions. The Court declined to review a companion case to *Wyatt v. Stickney* (*Aderholt*) [48]. It did consider *O'Connor v. Donaldson* [57] in 1975 but declined to see this case as a "right to treatment" case at all ([57], p. 573), although the lower courts had dealt with it in those terms ([57], note 6, pp. 570–572). The Supreme Court held, again in very general terms, "a State cannot confine without more a nondangerous individual who is capable of surviving safely in freedom by himself or with the help of willing and responsible family members or friends" ([57], p. 576). The Court in *Donaldson* held that the jury had had ample evidence from which to find a violation of this "right to freedom" ([57], p. 576).

The Court did not return to this central issue of a right to treatment for institutionalized persons again in any way until 1982. In a decision that appeared as this chapter was being completed, the Court continued its cautious approach, although it moved further toward a recognition of some enforceable rights for institutionalized retarded persons. In *Youngberg v. Romeo* the Court extended to the civil confinement area prior holdings from the penal confinement context to hold that there is a "right to safe conditions" and a "right to freedom from bodily restraint" that are constitutionally protected ([70], p. 316). The Court stated that "it must be unconstitutional to confine the involuntarily committed ... in unsafe conditions" ([70], p. 316). The Court also held that there is a right to "minimally adequate or reasonable training to ensure safety and freedom from undue restraint" ([70], p. 319). The Court interpreted the *Youngberg* case not to "present the difficult question whether a mentally retarded person, involuntarily committed to a state institution, has some general constitutional right to training *per se*, even when no type or amount of training would lead to freedom" ([70], p. 318).

If the Supreme Court has been cautious in recognition of a federal constitutional right to treatment or habilitation for persons within state institutions, it has been even less willing to set strict constitutional limits on the procedures by which people are committed to such institutions. *Lessard v. Schmidt* was the federal district court case, discussed in Part II above, where the District Court considered the constitutionality of Wisconsin's procedures for commitment of the mentally ill and held them to be seriously deficient [54]. Although the Supreme Court never addressed the substance of that case, it sent it back for reconsideration on collateral technical questions twice [54]. Then, the Supreme Court considered an appeal in another federal court case, where a federal district court had approved a commitment process that permitted the confinement of a person for as long as forty-five days before there was any opportunity for a hearing [46]. The Supreme Court affirmed that decision without even addressing the issues in a written opinion [46]. The net effect of these two cases is to indicate a very substantial tolerance by the Court in the procedural domain. In its one written opinion on a question of commitment procedure for adults, the Court has confirmed its tolerance [43]. In *Addington v. Texas*, the Court considered what standard of proof must be met in order validly to civilly commit a person [43]. The state argued for a "preponderance of the evidence" standard, which is utilized in the vast majority of non-criminal cases. The petitioner argued that because he was in danger of being confined, the "beyond a reasonable doubt" standard used in criminal trials should be required. This analysis had been embraced by the Court several years earlier for juvenile delinquency cases [52]. The Supreme Court held in *Addington*, however, that although more than the civil standard is required to sustain a civil commitment, less is required than for a criminal conviction. Using an analogy to deportation proceedings, which may say more than the Court intended about the nature of commitment, it prescribed a minimum standard of "clear and convincing evidence" ([43], p. 433), an intermediate standard.

The Supreme Court also has spoken to the procedural due process standards for commitment of children. In perhaps its weakest decision from a civil rights perspective, the Court held that there was no need for a judicial hearing at all in connection with the commitment of a child to a psychiatric hospital, even where the hospitalization is initiated by a county agency that is guardian to the child. It found adequate protection in professional and parental exercise of judgment [58].

Most interesting of all, however, are several cases involving the interpretation of federal statutes. It was noted at the end of Part II that early federal district court cases on the rights of the handicapped were quickly supplemented by federal statutes. Three key statutes in this process were the Education for All Handicapped Children Act of 1975 [72], the Developmentally Disabled Assistance and Bill of Rights Act [71], and Section 504 of the Rehabilitation Act Amendments of 1973 [73]. These federal statutes contained language reinforcing the rights to education and habilitation and the right to be free from discrimination. In the cases to be considered now, the Supreme Court was not asked to engage in constitutional interpretation or in "judicial legislation," or to create or expand constitutional rights. Rather, it had only to interpret and apply the language of the legislative branch. If the Supreme Court had interpreted congressional language as it had been interpreted by the Courts of Appeals, constitutional issues could have been avoided. Furthermore, Congress could have come back and amended the legislation if it found that the Court had gone further than Congress had contemplated. In three important decisions, the Supreme Court has given the narrower of two possible interpretations to the federal statute and has denied relief granted by lower courts. One can wonder whether these restrictive interpretations are part of a broader "backlash against the handicapped" [7].

The first of these cases is *Southeastern Community College v. Davis* [64]. Ms. Davis had been denied admission to a public nursing school program on the basis of a severe hearing impairment. Ms. Davis invoked the Section 504 of the Rehabilitation Act Amendments of 1973, which states:

> No otherwise qualified handicapped individual . . . shall, solely by reason of his handicap, be excluded from the participation in, or be denied the benefits of, or be subjected to discrimination under any program or activity receiving Federal financial assistance . . . [73].

Ms. Davis's argument was that a hearing impairment ought not to disqualify a person from training to be a nurse. Thus, she argued, exclusion from a nursing education program on the basis of such an impairment constituted a violation of this section. By implication, any impediments resulting from her hearing impairment ought to be compensated for. Impediments arise from the incongruity of two circumstances; the nursing program should adjust to permit hearing impaired persons to participate. Ms. Davis, after losing in the trial court, prevailed in the federal Court of Appeals. The United States Supreme Court, however, reversed [64]. It gave a restrictive interpretation to Section 504 of the Rehabilitation Act. By considering the existing comprehensive state licensing and training programs as legitimate, static elements, the Court characterized Ms. Davis's request as one for "affirmative action" rather than as one for removal of discriminatory barriers ([64], p. 407). The Court even paid lip service to the artificiality of the distinction it used. It said:

> We do not suggest that the line between a lawful refusal to extend affirmative action and illegal discrimination against handicapped persons always will be clear. It is possible to envision situations where an insistence on continuing past requirements and practices might arbitrarily deprive genuinely qualified handicapped persons of the opportunity to participate in a covered program. . . . [S]ituations may arise where a refusal to modify an existing program might become unreasonable and discriminatory ([64], pp. 412–13).

Nevertheless, the Court rejected the expansive view of Section 504 that had been embraced by the Court of Appeals.

The second case involving restrictive statutory interpretation is *Pennhurst State School and Hospital v. Halderman* [59]. In the *Pennhurst* case, as mentioned above, the federal District Court held that the conditions and programs (or lack thereof) at Pennhurst were in violation of plaintiffs' right to habilitation and right to habilitation in the least restrictive environment. The District Court held that these rights had many alternative bases: federal due process or equal protection; Section 504 of the Rehabilitation Act Amendments of 1973; and a state statute [49].

On appeal the Third Circuit agreed that there is a right to habilitation in the least restrictive setting [50]. It declined, however, to explore the District Court's various bases for the right, holding instead that the right is clearly provided by the Developmentally Disabled Assistance and Bill of Rights Act ([50], p. 97), whose provision is set forth in Part II of this chapter.

The Court of Appeals had noted the oft-repeated proposition that federal courts should use statutory bases for decisions to avoid constitutional issues where possible. It then had no difficulty in finding that Congress had articulated an enforceable right to habilitation in the least restrictive setting.

Justice Rehnquist, speaking for five members of a 6–3 majority of the Supreme Court, held that the above-quoted language creates no rights. He simply concluded that Congress had "no intention" to create rights by this language. He went on to say:

> It defies common sense, in short, to suppose that Congress implicitly imposed this massive obligation on participating states ([59], p. 24).

Suffice it to say that such an intent of Congress seems less nonsensical if Congress could have felt that the rights it articulated already burdened the states by virtue of the federal Constitution and if deinstitutionalization is seen as a national priority [12], ([32], p. 140). While the Supreme Court's argument is not absurd, the Court could quite reasonably have given effect to the Congressional language. Under such an approach, of course, Congress would have been free to reconsider the matter and change its law or augment its funding if it so wished. The Supreme Court made a choice.

Finally, as this chapter was in final preparation, the Court, in *Board of Education v. Rowley* [45], considered the third major federal statute that had been enacted during the civil rights decade for the handicapped. This was the Education for All Handicapped Children Act of 1975 [72], which was enacted in response to the two right to education cases discussed in Part II above ([45], p. 3043), the *P.A.R.C.* case [60] and *Mills v. Board of Education* [55]. The Act required states receiving federal funding for the education of handicapped children to have "a policy that assures all handicapped children the right to a free *appropriate* public education" ([72], § 1412(1) *emphasis supplied*). *Board of Education v. Rowley* [45] was the first case under the Act to be decided by the Supreme Court. *Rowley* addressed the issue of how one determines whether the public education provided to a handicapped child is "appropriate."

An "appropriate" public education is what the act requires ([72], § 1412(1)). In a six to three decision, with Mr. Justice Rehnquist again writing for the Court and five of its members, "appropriate" is given a very limited interpretation. Mr. Justice White, writing in dissent for himself and Justices Brennan and Marshall, says that "[i] n order to reach its result in this case, the majority opinion contradicts itself, the language of the statute, and the legislative history" ([45], p. 3053).

Amy Rowley was a deaf first grader. The school offered to teach Amy in a regular classroom, where she would use an FM hearing aid that compensated only partially for her hearing loss, and to provide three hours a week of speech therapy and an hour a day of tutorial instruction from a tutor for the deaf. Amy's parents, who were also deaf, agreed to these measures but insisted as well that a sign-language interpreter should be provided in all her academic classes. This the school refused to do ([45], p. 3039). The issue then became how to decide between the offer of the school administrators and the claim for an interpreter by the parents.

Following procedures provided by the statute, the parents pursued their claim before an "independent examiner" and the New York Commissioner of Education, each of whom affirmed the administrators' determination. The parents then sought judicial review in federal court, also as provided for in the statute ([45], p. 3040). The federal District Court [62] and Court of Appeals [63] reversed the administrative determination and ordered provision of an interpreter. The Supreme Court reversed these court decisions, reinstating the administrative determinations ([45], p. 3052). The importance of the case lies not in the particulars of Amy Rowley's handicap or the specifics of the remedy. The importance lies rather in the approach taken by the Court and its restrictive interpretation of the Education for All Handicapped Children Act of 1975.

The District Court found:

> Amy is a very bright child who is doing fairly well in school. . . . [S] he understands considerably less of what goes on in class than she could if she were not deaf. Thus she is not learning as much, or performing as well academically, as she would without her handicap ([62], p. 532).

The Court then went on to consider by what standard it should decide whether an interpreter was a necessary element of an "appropriate" education for her. It noted that the term "appropriate" is not defined, or at least not helpfully defined, in the act and then proceeded to discuss the alternative interpretations that could be given:

An "appropriate education" could mean an "adequate" education – that is, an education substantial enough to facilitate a child's progress from one grade to another and to enable him or her to earn a high school diploma. An "appropriate education" could also mean one which enables the handicapped child to achieve his or her full potential. Between those two extremes, however, is a standard which I conclude is more in keeping with the regulations, with the Equal Protection decisions which motivated the passage of the Act, and with common sense. This standard would require that each handicapped child be given an opportunity to achieve his full potential *commensurate with the opportunity provided to other children*. ... Since some handicapped children will undoubtedly have the intellectual ability to do better than merely progress from grade to grade, this standard requires something more than the "adequate" education described above. On the other hand, since even the best public schools lack the resources to enable every child to achieve his full potential, the standard would not require them to go so far ([62], p. 538, *emphasis supplied*).

The District Court concluded, in other words, that "appropriate" education is a relative concept requiring comparison of the handicapped child's aptitude with other children in class (relative capability) and comparison of the handicapped child's shortfall from her aptitude with that of other students. The District Court emphasized the school principal's repeated statement that "only [Amy's] academic failure would convince the school district that she needed the services of an interpreter" ([62], p. 534). It found avoidance of failure to be too low a standard to define the content of an "appropriate" education for a bright deaf child.

The Court of Appeals for the Second Circuit affirmed the District Court decision [63], although it went to some length to negate any precedential effect the case might have ([63], p. 948).

In the Supreme Court, Justice Rehnquist adopted for the Court the test for which the school principal had contended. The Court said:

Insofar as a State is required to provide a handicapped child with a "free appropriate public education," we hold that it satisfies this requirement by providing personalized instruction with sufficient support services to permit the child to benefit educationally from that instruction. Such instruction and services must be provided at public expense, must meet the State's educational standards, must approximate the grade levels used in the State's regular education, and must comport with the child's IEP [Individualized Educational Program]. In addition, the IEP, and therefore the personalized instruction should be formulated in accordance with the requirements of the Act and, if the child is being educated in the regular classrooms of the public education system, *should be reasonably calculated to enable the child to achieve passing marks and advance from grade to grade* ([45], p. 3049, *emphasis supplied*).

In each of these three cases involving interpretation of federal statutes enacted by Congress in support of the civil rights movement for the handicapped, the United States Supreme Court has restrictively interpreted broad remedial statutory language. When coupled with its caution in articulating constitutional protections for the handicapped, the Court must be seen to be playing a very constraining role.

College of Law
Ohio State University
Columbus, Ohio

BIBLIOGRAPHY

[1] American Bar Association, Commission on the Mentally Disabled: 1979, *Executive Summary, Exercising Judgment for the Disabled, Report of an Inquiry* (prepared by M. T. Axilbund).

[2] Allen, R.C., Ferster, E.Z., and Weihofen, H.: 1968, *Mental Impairment and Legal Incompetency*, Prentice-Hall, Englewood Cliffs, N.J.

[3] Baumeister, A.A.: 1970, 'The American Residential Institution: Its History and Character', in [4], pp. 1–28.

[4] Baumeister, A.A., and E. Butterfield (eds.): 1970, *Residential Facilities for the Mentally Retarded*, Aldine, Chicago.

[5] Birnbaum, M.: 1960, 'The Right to Treatment', *American Bar Association Journal* **46**, 499.

[6] Brakel, S.J. and Rock, R.S. (eds.): 1971, *The Mentally Disabled and the Law, An American Bar Foundation Study* (Revised Edition), The University of Chicago Press, Chicago.

[7] Breslin, E.R.: 1980, 'Backlash Against the Disabled', *Mental Disability Law Reporter* **4**, 345.

[8] Brief for Plaintiff in Error in *Buck v. Bell*, 274 U.S. 200 (1927) in P. B. Kurland, and G. Casper (eds.): 1975, *Landmark Briefs and Arguments of the Supreme Court of the United States: Constitutional Law*, Volume 25, University Publications of America, Inc., Arlington, Va., pp. 491–509.

[9] Burgdorf, R.L. and Burgdorf, M.P.: 1977, 'The Wicked Witch is Almost Dead: *Buck v. Bell* and the Sterilization of Handicapped Persons', *Temple Law Quarterly* **50**, 995 (*reprinted in* [31], pp. 995–1034).

[10] Burgdorf, R.L. (ed.): 1980, *The Legal Rights of Handicapped Persons*, Paul H. Brookes, Publishers, Baltimore.

[11] Burt, R. and Morris, N.: 1972, 'A Proposal for the Abolition of the Incompetency Plea', *Univ. of Chicago Law Review* **40**, 66.

[12] Comptroller General of the United States: 1976, *Returning the Mentally Disabled to the Community: Government Needs to Do More*, U.S. Govt. Printing Office, Washington.

[13] Cox, I.W.: 1934, 'The Folly of Human Sterilization', *Scientific American* **151**, 188.

[14] Cynkar, R.J.: 1981, '*Buck v. Bell*: "Felt Necessities" v. Fundamental Values?', *Columbia Law Review* **81**, 1418.

[15] Dybwad, G.: 1964, *Challenges in Mental Retardation*, Columbia, New York.

[16] Ferster, E.: 1966, 'Eliminating the Unfit – Is Sterilization the Answer?' *Ohio State Law Journal* **27**, 591.

[17] Foote, C.: 1960 'A Comment on Pre-trial Commitment of Criminal Defendants', *Univ. of Pennsylvania Law Review* **108**, 832.

[18] Gilhool, T.K.: 1976, 'The Right to Community Services', in M. Kindred, J. Cohen, D. Penrod, and T. Shaffer (eds.), *The Mentally Retarded Citizen and the Law*, Free Press, New York, pp. 173–207.

[19] Gosney, E.S.: 1934, 'Eugenic Sterilization – Human Betterment Demands It', *Scientific American* **151**, 18.

[20] Herr, S.: 1976, 'The Right to an Appropriate Free Public Education', in M. Kindred, J. Cohen, D. Penrod and T. Shaffer (eds.), *The Mentally Retarded Citizen and the Law*, Free Press, New York, pp. 252–267.

[21] Kanner, L.: 1964, *A History of the Care and Study of the Mentally Retarded*, Charles C. Thomas, Springfield, Illinois.

[22] Kindred, M., J. Cohen, D. Penrod, and T. Shaffer (eds.): 1976, *The Mentally Retarded Citizen and the Law*, Free Press, New York.

[23] Kindergan, C.P.: 1966, 'Sixty Years of Compulsory Eugenic Sterilization: "Three Generations of Imbeciles" and the Constitution of the United States', *Chicago-Kent Law Review* **43**, 123.

[24] Landman, J.H.: 1934, 'Race Betterment by Human Sterilization', *Scientific American* **150**, 292.

[25] Laughlin, H.H.: 1922, *Eugenical Sterilization in the United States*, Psychopathic Laboratory of the Municipal Court of Chicago, Chicago, Illinois.

[26] Lippman, L. and Goldberg, I.I.: 1973, *Right to Education*, Teachers College Press, N.Y.

[27] Note: 1967, 'Civil Restraint, Mental Illness, and the Right to Treatment', *Yale Law Journal* 77, 87.

[28] Note: 1967, 'Incompetency to Stand Trial', *Harvard Law Review* **81**, 454.

[29] Note: 1976, 'The Education of (sic) All Handicapped Children Act of 1975', *Univ. of Michigan Journal of Law Reform* **10**, 110.

[30] Note: 1979, 'Enforcing the Right to an "Appropriate" Education: The Education for All Handicapped Children Act of 1975', *Harvard Law Review* **92**, 1103.

[31] Phillips, W.R.F. and Rosenberg, J. (eds.): 1980, *Changing Patterns of Law: The Courts and the Handicapped*, Arno Press, N.Y.

[32] President's Committee on Mental Retardation: 1976, *Mental Retardation: Century of Decision*, U.S. Govt. Printing Office, Washington.

[33] President's Committee on Mental Retardation: 1977, *Mental Retardation: Past and Present*, U.S. Govt. Printing Office, Washington.

[34] President's Panel on Mental Retardation: 1963, *A National Plan to Combat Mental Retardation*, U.S. Govt. Printing Office, Washington.

[35] Roos, P.: 1970, 'Evolutionary Changes of the Residential Facility', in Baumeister, A.A. and E. Butterfield (eds.), *Residential Facilities for the Mentally Retarded*, Aldine, Chicago, pp. 29–58.

[36] Roos, P.: 1976, 'Basic Personal and Civil Rights – Reaction Comment', in M. Kindred, J. Cohen, D. Penrod, and T. Shaffer (eds.), *The Mentally Retarded Citizen and the Law*, Free Press, New York, pp. 26–28.

[37] Rosen, M., Clark, G.R., and Kivitz, M.S. (eds.): 1976, *The History of Mental Retardation* (2 vols.), University Park Press, Baltimore.

[38] Rothman, D.: 1976, 'The Right to Habilitation – Reaction Comment', in M. Kindred, J. Cohen, D. Penrod, and T. Shaffer (eds.), *The Mentally Retarded Citizen and the Law*, Free Press, New York, pp. 407–411.

[39] Symposium: 1969, 'The Right to Treatment', *Georgetown Law Journal* 57, 673.

[40] Thomalla, C.: 1934, 'The Sterilization Law in Germany', *Scientific American* **151**, 126.

[41] Vukowich, W.T.: 1971, 'The Dawning of the Brave New World – Legal, Ethical, and Social Issues of Eugenics', *University of Illinois Law Forum* 189.

[42] Wolfensberger, W.: 1975, *The Origin and Nature of our Institutional Models*, Human Policy Press, Syracuse, New York.

[43] *Addington v. Texas*, 441 U.S. 418 (1979).

[44] *Baxstrom v. Herold*, 383 U.S. 107 (1966).

[45] *Board of Education of the Hendrick Hudson Central School District v. Rowley*, 102 S. Ct. 3034 (1982).

[46] *Briggs v. Arafeh*, 411 U.S. 911 (1973), *affirming by memorandum Logan v. Arafeh*, 346 F. Supp. 1265 (D. Conn. 1972).

[47] *Buck v. Bell*, 274 U.S. 200 (1927).

[48] *Dept. of Human Resources of Georgia v. Burnham*, 422 U.S. 1057 (1975), *denying certiorari from* 503 F.2d 1319 (5th Cir. 1974).

[49] *Halderman v. Pennhurst State School and Hospital*, 446 F. Supp. 1295 (E.D.Pa. 1977).

[50] *Halderman v. Pennhurst State School and Hospital*, 612 F.2d 84 (3d Cir. 1979).

[51] *Humphrey v. Cady*, 405 U.S. 504 (1972).

[52] *In re Winship*, 397 U.S. 358 (1970).

[53] *Jackson v. Indiana*, 406 U.S. 715 (1972).

[54] *Lessard v. Schmidt*, 349 F. Supp. 1078 (E.D. Wisc. 1972); *vacated and remanded* 414 U.S. 473 (1973); 379 F. Supp. 1376 (1974); *vacated and remanded* 421 U.S. 957 (1975); 413 F. Supp. 1318 (1976), *prior judgment reinstated*.

[55] *Mills v. Board of Education of the District of Columbia*, 348 F. Supp. 866 (D.D.C. 1972).

[56] *Nason v. Superintendent, Bridgewater State Hospital*, 353 Mass. 604, 233 N.E.2d 908 (1968).

[57] *O'Connor v. Donaldson*, 422 U.S. 563 (1975).

[58] *Parham v. J.R.*, 442 U.S. 584 (1979).

[59] *Pennhurst State School and Hospital v. Halderman*, 451 U.S. 1 (1981).

[60] *Pennsylvania Association for Retarded Children v. Pennsylvania*, 334 F. Supp. 1257 (E.D.Pa. 1971), *modified* 343 F. Supp. 279 (1972).

[61] *Rouse v. Cameron*, 373 F.2d 451 (D.C.Cir. 1966).

[62] *Rowley v. Board of Education of Hendrick Hudson Central School District*, 483 F. Supp. 528 (S.D.N.Y. 1980).

[63] *Rowley v. Board of Education of Hendrick Hudson Central School District*, 632 F.2d 945 (2d Cir. 1980).

[64] *Southeastern Community College v. Davis*, 442 U.S. 397 (1979).
[65] *Specht v. Patterson*, 386 U.S. 605 (1967).
[66] *Stump v. Sparkman*, 435 U.S. 349 (1978).
[67] *Wyatt v. Stickney*, 325 F. Supp. 781 (M.D.Ala. 1971), *modified* 344 F. Supp. 373 (1972).
[68] *Wyatt v. Stickney*, 344 F. Supp. 387 (M.D.Ala. 1972).
[69] *Wyatt v. Aderholt*, 503 F.2d 1305 (5th Cir. 1974).
[70] *Youngberg v. Romeo*, 457 U.S. 307 (1982).
[71] Developmentally Disabled Assistance and Bill of Rights Act, Pub. L. No. 94–103, 89 Stat. 486 (1975), codified at 42 U.S.C. §§ 6001–6081 (1976).
[72] Education for all Handicapped Children Act of 1975, Public Law No. 94–142, 89 Statutes 773 (1975), codified at 20 U.S.C. §§ 1401–1461 (1976).
[73] Rehabilitation Act of 1973, § 504, codified at 29 U.S.C. § 794 (1976).
[74] Ohio Revised Code Annotated (Page 1966).
[75] Virginia Acts of Assembly, 1924, pp. 569–571.

BARBARA BAUM LEVENBOOK

EXAMINING LEGAL RESTRICTIONS ON THE RETARDED

Discussions of the legal rights of the retarded have revealed that retarded people typically have both special privileges and special restrictions in the law. Among the privileges are a legal incompetent's immunity in contract which, as Glanville Williams has pointed out, is really a liberty not to pay what would otherwise be his contractual debts.[1] (Whether that liberty is really an advantage, which is what the term 'privilege' usually suggests, is another question.) Among the restrictions in most states is the denial to a legal incompetent of a right to marry. At some time or other, retarded people have been denied the legal right to vote, to decide whether and when to have children, to serve on juries, and the right as children to a free public education. Some of these restrictions are commonplace today. Such legal restrictions have often been criticized,[2] or defended,[3] by courts and legal writers without an appreciation of the complexity of the moral issues they raise. My purpose in this paper is to examine various assumptions one might make about the moral status of the retarded in order to support a conclusion that they should have certain legal restrictions. The moral arguments for certain legal restrictions are more complex than has been appreciated by proponents or opponents of restrictions in the legal literature.

Legal restrictions may take various forms, and may be subtly interwoven with accompanying legal privileges (like immunity in contract).[4] Their net effect, however, is to reduce the legal status of the persons restricted in comparison with the legal status of normal adult citizens. A person who is legally restricted may lose what Hohfeld called a legal liberty (thus gaining a legal duty that others do not have), or he may lose an immunity to the exercise of power by others.[5] More typically, when a retarded person loses a right, what the person loses are, in Hohfeldian terms, powers he would otherwise have to create duties for others. Usually, in addition, duties on the part of others that serve to protect his legal liberties are abrogated.[6] Those duties might be duties not to interfere with his exercise of liberty, duties to provide something or do something on his behalf, or combinations of the two. Indeed, the law may give others a duty to act in ways inconsistent with his exercise of liberty, as when it requires a guardian to make certain decisions for the legal incompetent. In sum, one who loses a legal liberty may also lose

L. Kopelman and J. C. Moskop (eds.), Ethics and Mental Retardation, 209–221.

the protection of that liberty that resides in the duties, powers, and immunities of others. For simplicity, I will follow ordinary usage and speak of legal restrictions as the denial of legal rights; but it should be borne in mind that the expression 'the denial of legal rights' covers a multitude of sins.

From what assumptions might one start in defending or criticizing the denial of legal rights to the retarded? One of the fundamental assumptions is that retarded persons have interests. Without this assumption, appeals to paternalism – either for special restrictions or for special privileges – are out of the question. For if someone does not think a retarded person is the sort of being that has had, does have, or will have interests, he must hold that a retarded person cannot have a good, and cannot be harmed.[7] Hence, nothing can be done for his good, or to prevent harm to him. Paternalistic justifications of treatment of him would be as inappropriate as paternalistic justifications of the treatment of rocks or public monuments. Just as we "protect" rocks and public monuments for the sake of those who have interests (in their preservation), so, too, if the retarded do not have interests, we can "protect" them only for someone else's sake.

My own view is that the case for retarded people – even the most profoundly retarded – having interests is as strong as the case for animals or young infants having interests. Hence, the burden of proof is on anyone who would argue otherwise. (This may help to explain why many critics of legal restrictions on the retarded simply assume that the retarded have interests.[8])

Some philosophers (notably Feinberg) believe that possession of interests is a necessary condition for possession of moral rights,[9] and this brings us to the next fundamental assumption about the retarded. It is sometimes assumed that the retarded have moral rights – especially rights to freedom and self-determination.[10] (Freedom here ought not to be confused with Hohfeldian liberty, which is the absence of a duty.) Without the assumption that the retarded have such rights, it is questionable that there is any strong moral presumption against paternalistic interferences with their choices and decisions, as when they are institutionalized against their will. General utilitarian considerations may give rise to presumptions against interfering with someone's liberty or autonomy, but these presumptions are no weightier than a case that can be made on general utilitarian principles for interference with special groups of people, like the retarded. Perhaps considerations of humanity or distributive justice (if the latter is not restricted to rights-bearers) give rise to a presumption against interfering with the choices and decisions of the retarded in some circumstances. However, when the interference is aimed at securing their good, humanity would seem to be on the side of the interference. Distributive justice might seem a better candidate for a value that gives rise to a general presumption against interference, but it is limited in the non-ideal world to social distributions that are feasible. So, if under one scheme the retarded are routinely interfered with and if all feasible alternatives are as unjust or more so (even though the retarded are less interfered with), then distributive justice cannot ground an objection to continuing with the present scheme. In contrast, a rights view can.

Do the retarded have rights to freedom and self-determination? If retarded persons possess the complex set of abilities necessary for autonomy (whatever disparate abilities this involves), it would seem that the burden of proof is on someone who argues that they do not. These abilities, however, which one might sum up as the ability to choose for oneself, are not all-or-nothing things. One may have any one of them to a degree, or have it with respect to some possible choices and not others. When one has partial abilities, it is not clear what should be said about a concomitant right to self-determinaton. Perhaps one can have a right to self-determination limited to certain decisions. But can one have a moral right to self-determination that is only partial with respect to all decisions? What would such a right be like?

Perhaps a case can be made that those without the ability to choose for themselves do not have a right to self-determination. (If, however, there were a way of making such people partially able to choose – through some therapy, for example, that increased I.Q. – they might have a right to be made able to choose, that is, a right to be made the sort of being that has a right to self-determination.) But even when a retarded person has a right to freedom or self-determination, it may be possible to justify restricting his freedom or infringing his autonomy through denial of legal rights. What is required is that the justification given is either compatible with respecting his rights or strong enough to outweigh them.

Ordinarily, appeals to beneficence will not be.[11] One cannot, for example, justify coercively preventing a rational, competent adult from marrying his intended on the grounds that he will be benefited by this interference or protected from the unhappiness of a predictable divorce; and the reason seems to be that rational, competent adults have rights to self-determination that are weightier than the considerations of beneficence adduced against them. Similarly, consider what would be said against a proposal by the exceptionally gifted to deny voting rights to the intellectually normal on the grounds that most of the normal do not fully comprehend all the risks to their well-being they run by their voting habits. Surely, the reply would be that the right to self-determination possessed by the intellectually normal includes the right to choose unwisely.

However, under certain conditions, the individual's own welfare, in combination with other factors, may be either compatible with respecting the right to self-determination or strong enough to outweigh it. On this point, there is general agreement among philosophers. But there is less agreement about what those conditions are. Some philosophers regard any paternalistic intervention as justified when it protects the individual from serious harm and when his decision to risk that harm has certain less-than-ideal features. On one view proposed by Dworkin [1], paternalism is justified only if the decision is less than fully rational. Dworkin's proposal, however, has been criticized for permitting too much paternalistic intervention in the lives of each of us ([17], pp. 198–199); almost none of the decisions of even the most intellectually gifted of us are *fully* rational. Similarly, C. L. Ten's view that paternalism is justified whenever a person decides under conditions of serious impairment is flawed, for it permits too much paternalistic intervention even when applied to persons of normal intellectual abilities (whose decisions are, for instance, temporarily impaired by alcohol).[12] Feinberg [3] argues that paternalism is justified only when a decision is substantially nonvoluntary. Feinberg's view, however, will not justify denying rights to all or even most of the retarded; since retarded people, like other people, do not usually make substantially nonvoluntary decisions. The intellectual capacities the retarded lack have little connection with the possibility of voluntary decision-making.

Feinberg is one of several philosophers who hold that paternalism is only justified when it avoids infringing the right to self-determination (sometimes referred to as the right to autonomy).[13] Noninfringement of this right is usually explained in terms of the individual's actual or hypothetical consent to the infringement. It seems obvious that actual consent by the retarded to legal restrictions is unlikely. So to apply such a view, we have to examine what is meant by hypothetical consent to a legal restriction.

Except for one philosopher who interprets hypothetical consent in terms of what hypothetical persons – namely, fully rational individuals – would prefer,[14] most philosophers who base the justification of paternalism on the possibility of hypothetical consent mean the consent of the person restricted. But usually the test is not whether he would consent, given his preferences at the time of the proposed restriction. For paternalistic restrictions are sought especially when someone's choices and decisions are in the same way impaired or encumbered – as, for instance, they arguably are when he is retarded, mentally ill, drunk, hysterical, or in ignorance of something relevant to his decision. Under these conditions, some of his preferences – especially, his desires concerning being restricted by others – are likely to be due to the condition that impairs or encumbers his decision-making. The test is rather whether he, the person restricted, would consent if he were not impaired or encumbered. Such a test, by the way, is analogous to the one by which courts measure the permissibility of treatment of persons regarded as mentally incompetent. Using what is called the 'principle of substituted judgment', courts regard themselves as obligated to determine, as far as is possible, what the incompetent person would want if he were competent.[15]

Two comments should be made about this interpretation of hypothetical consent. First, if the justification of paternalistic restriction rests upon it, then the denial of legal rights cannot be paternalistically justified in the case of the very severely or profoundly retarded. This is because the principle of hypothetical consent is inapplicable to someone who is so severely retarded that his personal identity is bound up in his impairing condition. For instance, in the well-known case of *Superintendent of Belchertown v. Saikewicz* [24], Joseph Saikewicz was an institutionalized sixty-seven year old man with an I.Q. of ten and a mental age of two. The decision before the Massachusetts Supreme Court was whether or not to provide for him a course of chemotherapy designed to retard, but not cure, the leukemia from which he was dying. According to the principle of substituted judgment as I have presented it, the issue for the court was what an intellectually normal Joseph Saikewicz, faced with the same medical decision, would prefer.[16] But an intellectually normal Joseph Saikewicz is in crucial respects indeterminate. Some normal people prefer to risk the side effects of chemotherapy in order to prolong their lives and some do not; and there is nothing in the nature of Saikewicz's case to make one preference more appropriate to assign to a hypothetical "him" than another. This is *not* an epistemological point about the limits of our knowledge. The real Joseph Saikewicz, being so very abnormal, has neither preference nor is capable of either preference, nor is committed in his preferences or values to anything that would determine what his preference would be if he were normal. Indeed, it seems to be doubtful that it would be *he* who was normal, since nearly everthing about him that has been asserted to be relevant to personal identity – personality, character traits, and even, possibly, his brain – would have to be different for him to be intellectually normal.

Second, it is doubtful that legal restrictions on even the mildly or moderately retarded could be justified by such a paternalistic principle. I have in mind restrictions like denial of the right to marry, to decide whether or not to have children, or to decide on treatment (where available). (For other restrictions – denial of the right to vote, for instance, or denial of the right to serve on juries – a paternalistic justification may not be appropriate. For it is difficult to see how it is ordinarily for the good of the person denied these rights that he is denied them – as opposed, for example, to being for the good of the community in general.) One reason why denials to the mildly or moderately retarded of a right to, for instance, decide on treatment is unlikely to be paternalistically justified is that mild or moderate retardation itself is not necessarily an impairment to such a decision. (More on this below.) And if this is so, then an autonomy-based paternalistic principle must make the justification of legal restrictions depend on whether those unimpaired retarded individuals would, if asked, consent to the legal restriction. Chances are, they would not.

But the more important reason is that even when mild or moderate retardation is an impairment to making decisions of the sort in question, it is not always true that the retarded person, if intellectually normal, would consent to the legal restriction. Even intellectually normal people differ in the weight they give the freedom to live their lives according to their own lights, compared to the benefits of being legally protected from making foolish or even highly imprudent decisions. There are some normal people – particularly some philosophers who write about paternalism – who value the former so much that they would consent only to restrictions designed to save their lives or their freedom, particularly when they know that these restrictions will be enforced with all the power of the legal system and administered in procedures that might invite discretionary abuse.[17]

What this altogether-too-brief discussion of paternalism and the retarded suggests, then, is that on currently defended principles of paternalism, legal restrictions of the mildly or moderately retarded are unlikely to be paternalistically justified. But it would be premature to conclude that legal restrictions of the mildly or moderately retarded are not justified at all. Paternalism is not the only possible justification. Although space does not permit me to discuss all of them, I would like to close by examining one other possible justification for legally restricting the retarded: distributive justice.

Distributive justice is often thought of as a reason for extending special legal rights *to* the retarded – e.g., a right to appropriate education, to treatment when possible, and so on. However, it has been used as a justification for restricting, in the sense of disabling (by removing the power and protections of powers of), those not fully competent.[18] In order to make use of such an argument against the retarded, it should be pointed out, yet another assumption about their moral status must be made: namely, that they are subjects of justice. Such an assumption is controversial. On Rawls' view, for instance, principles of justice apply only to those beings capable of a sense of justice and a conception of their good expressed in a rational plan of life;[19] so on Rawls' view, the question of whether the retarded are subjects of justice depends upon whether or not they are so capable. It might turn out that some of them are not (e.g., the severely or profoundly retarded). If so, and if Rawls is right about the scope of justice, then the severely and profoundly retarded cannot be unjustly discriminated against or treated unfairly. (The Supreme Court in *Jackson v. Indiana* [23, p. 733], however, was of another opinion. They quoted with approval language of another court to the effect that coercively incarcerating a severely retarded person incompetent to stand trial was substantially unjust.) I will either restrict the discussion to those retarded capable of a sense of justice and a conception of the good or suppose that Rawls is wrong about the beings to which principles of justice extend.[20]

The argument for the legal restriction of the retarded would go, then, as follows: Justice requires that each person have an equal opportunity to live the best life of which he or she is capable. In making certain decisions, the retarded have a handicap that prevents them from having a chance equal to that of others. By requiring those competent to make these decisions to choose for the retarded on the basis of what is best for them, we make their chances for a happy life more nearly equal to others'. When the only practical way of insuring that this is done is to legally disable the retarded from making these decisions, we are justified in disabling them.

Such an argument strengthens the case for paternalistic restriction of the retarded by an appeal to what is required by substantive justice. In other words, it claims that paternalistic acts may be required by justice, not merely permitted on the grounds of beneficence. The strategy is an interesting one; it rests on the assumption that when justice considerations are adduced on behalf of legal restriction, the case against restriction (where such a case exists) – namely, that it violates the right to freedom or autonomy – is outweighed. However, whether the argument is strong enough to outweigh the case against restriction is not entirely obvious. More would have to be said about the relative value of the right to autonomy, although the burden, it seems to me, is once again on anyone arguing for legal restrictions, rather than against them.

Enough can be said on other grounds, however, to undermine the appeal of the argument. That is, some of the presumptions of the argument are questionable. For example, one of the things that makes a life good is the self-esteem that comes from being self-directed, from making and being allowed to carry out one's own choices. This argument presupposes that the loss in the ability to make his own choices effective suffered by the retarded person who is legally restricted (and is capable of autonomy) will be more than compensated for in the preservation of chances for other things that make life good. Why, then, do we not say the same about restrictions on persons of normal intelligence, who may also make decisions that unwisely reduce their chances of having a good life – e.g., the woman who refuses to have a timely operation on her endometriosis, even though this will foreclose the opportunity to bear a child, which she very much wants?

One reason we do not is because any defensible equal opportunity principle of distributive justice must count chances or opportunities in such a way that they are unaffected by the individual decisions of adults about their own lives. If not, if we allow that decisions such as the woman's can cause injustice in opportunity-distributions, then justice would require us interfere too often with the decisions of adults. Justice would require continual readjustments of distributions of opportunities after a number of individual decisions or it would require coercively restraining people from choosing in ways that necessitate these readjustments. This is certainly implausible.

I suggest, then, that *if* justice requires that each person have an equal opportunity to live the best life of which he or she is capable, that requirement must be interpreted in such a way as to be compatible with the freedom of normal adults to make decisions concerning medical treatment for themselves, their education, and so on.[21] And if this is so, why would an unjust distribution of opportunities result if a retarded adult is permitted to carry out a decision that unwisely closes off opportunities – e.g., if he refuses to undergo an operation that will "cure" him of retardation[22] or unwisely refuses to enroll in a special night-school class that will t4ach him to read?

At this point, the proponent of legal restriction may switch tactics. Instead of arguing that restrictions are required by distributive justice, he may argue that they are compatible with it. He might take this line: Distributive justice requires that each person have an equal chance for the best life of which he or she is capable, but it does not require an equal chance to do things for which some are not capable and others are. It is not distributively unjust to close off roles (such as brain surgeon or jet pilot) to those normal adults who are not competent for them. So, too, it is not distributively unjust to close off roles (such as voter, jury member, married person, or decider-of-one's-own medical treatment) to those retarded adults who are not competent for them.

This brings us back to the question of whether or not retarded persons possess the capacity for autonomy, upon which a right to self-determination depends. For, in order to close off basic roles to all of the retarded, some factual assumptions must be made that are at least questionable. For instance, it must be assumed that all retarded persons, or most of them anyway, will be incompetent as decision-makers when faced with choices like that mentioned above. This assumption is questionable not only because retarded people are a widely diverse group but also because competence, like the more general capacity for autonomy, is not an all-or-nothing thing. In decision-making, it can be analyzed into a number of distinct abilities – e.g., the ability to distinguish relevant from irrelevant considerations, the ability to weigh such considerations against each other, the ability to remember the results of one's weighing long enough to act on them and so on. (Competence also requires something that is not an ability at all, though it presupposes abilities: knowledge of a sufficient number of relevant considerations.) People rarely lack all of these abilities or any one of them entirely. More important, a deficiency that someone has in one of these abilities may manifest itself only in some decisions, but not in others. One can be incompetent to decide whether the U.S. should return to the gold standard but be competent to decide how to budget a weekly paycheck. A retarded person incompetent to follow the intricate arguments in a tax case may be competent as a jury member in a simple assault case.

Lastly, it is doubtful that discrimination in legal rights on the grounds of a minority's incompetence for their exercise is just when that incompetence is socially induced. Certainly discrimination for social roles may not be just *because* incompetence is socially induced. For instance, as I have pointed out elsewhere [10, p. 250], in our own society just after World War II many women and blacks were incompetent for the roles of doctor and lawyer. For not only were they generally not admitted to professional schools (thus losing the opportunity to become training-competent for these roles), but their admission was often denied on the grounds of their predicted inability to function effectively in such roles. They may well have been ineffective – in that social mileu. Distrust and disrespect of blacks and women seem to have been so widespread that they might have had difficulty forming a client population, even among their own kind. (A prejudice against a group can be so pervasive that it affects the way members of that group judge their own.) There may have been insurmountable obstacles to women's or blacks' forming the sort of professional contacts with their influential and well-established colleagues (most of them white and male) that are necessary for success in the field. Nonetheless, in spite of the (educational) incompetence of women and blacks for these roles, I think we would all agree that the discrimination against women and blacks associated with these roles was unjust, and unjust because the incompetence was, at heart, caused by the social mileu in the first place.

Daniel Wikler [20] has recently suggested that much of the incompetence of mildly retarded persons – e.g., their incompetence to understand contractual obligations – is similarly socially induced. It is induced by the standard set by the demands of the society. A society might restructure its institutions so that retarded people (or more plausibly, the mildly retarded) would no longer be incompetent to function in them. For example, contracts could be void when misunderstood by one of the parties or when one of the parties has not been informed, in terms which he can understand, how his contractual obligations may inconvenience him. Marital obligations and their connection with economic support or responsibility for children might be revamped. The point is that insofar as Wikler is correct, the incompetence of the retarded for some decision-making cannot be easily reconciled with the demands of distributive justice for equal opportunities. Whether that reconciliation can be achieved depends upon further considerations, among them, as he points out, the question of the just distribution of the costs of restructuring society to minimize this sort of socially-created incompetence. That examination will have to await another forum.

North Carolina State University
Raleigh, North Carolina

NOTES

[1] Glanville Williams [21], p. 127. This immunity has exceptions and often applies only to those who are adjudged to be incapable of comprehending the subject, nature, and consequences of the contract. See Laurence P. Simpson [15], pp. 234–236.

[2] For a widespread condemnation of legal restrictions, see Patricia M. Wald [18].

[3] See, e.g., Justice Holmes' decision in *Buck v. Bell* [22].

[4] For instance, once a guardian is legally appointed for a person, he loses the power to prevent the guardian from disposing of his property as the guardian sees fit.

[5] For explanations of the terms 'liberty', 'duty', 'power', and their relationships, see Wesley Newcomb Hohfeld [6], especially pp. 35–64.

[6] According to Hohfeld [6], A loses a legal right against another, B, when and only

when B loses a duty to A, and when the right and the duty in question have the same content (e.g., a right to noninterference and a duty of noninterference). I am departing from his use of 'right'. Carl Wellman has defended this departure [19], calling legal rights "complex structure[s] of legal liberties, claims, powers and immunities" on the part of the rightholder, second parties, third parties, and officials (p. 52).

7 For an argument that A's presently having interests is not necessary for A's being harmed, see [11].

8 See, for example, Eleanor S. Elkin [2] (the profoundly retarded have interests which can conflict with the interests of their guardians).

9 Joel Feinberg [4]. I am inclined to disagree, but my reasons cannot be elaborated here.

10 See, e.g., Daniel Wikler [20] for a claim that the mildly retarded have a right to self-determination.

11 Nor will appeals to what benefits the rest of the adult population be sufficient. One such appeal is the basis of Justic Holmes' decision upholding the liberty of institutions to sterilize retarded patients in *Buck v. Bell* [22].

12 Point by Hodson [5].

13 See, for example, Donald VanDeVeer [17] and John D. Hodson [5].

14 Dworkin [1]. VanDeVeer [17] offers a detailed examination of what he calls Dworkin's 'Hypothetical Rational Consent' principle.

15 For a discussion of the principle of substituted judgment, see Gary E. Jones [8]. Note, however, that courts do not decide whether the incompetent would consent to the interference with his wishes, but rather whether he would agree with the choice being made for him. Presumably, the former entails the latter, but not vice versa.

16 As Jones [8] points out, the substituted judgment test has sometimes been given a different form by the courts. It is sometimes put in terms of what a rational person would choose in the circumstances of the legal incompetent.

17 John Stuart Mill, for instance, is associated with the view that society ought not to interfere paternalistically with the decisions of generally rational and competent adults except to prevent them from selling themselves into slavery or otherwise forfeiting their freedom ([12], p. 125). Jeffrie G. Murphy [13], pp. 484–485, notes that rational agents might be reluctant to agree to paternalistic interventions because of the potential for abuse of any paternalistic scheme administered through law.

18 Laurence D. Houlgate [7] has used distributive justice as a justification for restricting the rights of children. I have modeled the argument for restricting the rights of the retarded on his argument.

19 John Rawls [14]. Rawls explicitly says only that this capacity is sufficient for being entitled to justice, but he is tempted to say that it is necessary as well.

20 For an argument that Rawls is arbitrary in his restriction of principles of justice to beings with certain capacities see Donald VanDeVeer [16].

21 I limit this principle to adults because children involve special considerations. Depending on how life opportunities are to be measured, justice may require restricting children's freedom so as to hold open opportunities for them in the future.

22 Such an operation is envisioned in a short story by Daniel Keyes [9].

BIBLIOGRAPHY

[1] Dworkin, G.: 1971, 'Paternalism', in R. Wasserstrom (ed.), *Morality and the Law*, Wadsworth Publishing Company, Belmont, California, pp. 107–126.

[2] Elkin, E.S.: 1976, 'Reaction Comment', in M. Kindred *et al.* (eds.), *The Mentally Retarded Citizen and the Law*, Free Press, New York, pp. 87–92.

[3] Feinberg, J.: 1971, 'Legal Paternalism', *Canadian Journal of Philosophy* **1**, 105–124.

[4] Feinberg, J.: 1974, 'The Rights of Animals and Unborn Generations', in W. T. Blackstone (ed.), *Philosophy and Environmental Crisis*, University of Georgia Press, Athens, Georgia, pp. 43–68.

[5] Hodson, J.D.: 1977, 'The Principle of Paternalism', *American Philosophical Quarterly* **14**, 61–69.

[6] Hohfeld, W.D.: 1964, *Fundamental Legal Conceptions as Applied to Judicial Reasoning*, Yale University Press, New Haven, Connecticut (originally published 1919).

[7] Houlgate, L.D.: 1979, 'Children, Paternalism, and Rights to Liberty', in O. O'Neill and W. Ruddick (eds.), *Having Children*, Oxford University Press, New York, pp. 265–278.

[8] Jones, G.E.: 1980, 'The Principle of Substituted Judgment and the Treatment of Mental Incompetents', paper presented to the Annual Western Division Meeting of the American Philosophical Association, Denver Hilton Hotel, Denver, Colorado.

[9] Keyes, D.: 1964, 'Flowers for Algernon', in L. del Rey (ed.), *The Science Fiction Hall of Fame*, Vol. 1, Avon Books, New York, pp. 605–635 (originally published 1959).

[10] Levenbook, B.B.: 1980, 'On Preferential Admission', *Journal of Value Inquiry* **14**, 255–273.

[11] Levenbook, B.B.: 1981, 'Harming Someone After His Death', unpublished.

[12] Mill, J.S.: 1956, *On Liberty*, Bobbs-Merrill Company, Indianapolis (originally published 1859).

[13] Murphy, J.G.: 1974, 'Incompetence and Paternalism', *Archiv für Rechts- und Sozialphilosophie* **60**, 465–486.

[14] Rawls, J.: 1971, *A Theory of Justice*, Harvard University Press, Cambridge, Massachusetts.

[15] Simpson, L.P.: 1965, *Handbook of the Law of Contracts*, 2nd ed., West Publishing Company, St. Paul, Minnesota.

[16] VanDeVeer, D.: 1979, 'Of Beasts, Persons, and the Original Position', *Monist* **62**, 368–377.

[17] VanDeVeer, D.: 1980, 'Autonomy Respecting Paternalism', *Social Theory and Practice* **6**, 187–207.

[18] Wald, P.M.: 1976, 'Basic Personal and Civil Rights', in M. Kindred *et al.* (eds.), *The Mentally Retarded Citizen and the Law*, Free Press, New York, pp. 3–26.

[19] Wellman, C.: 1975, 'Upholding Legal Rights', *Ethics* **86**, 49–60.

[20] Wikler, D.: 1979, 'Paternalism and the Mildly Retarded', *Philosophy and Public Affairs* **8**, 377–392.

[21] Williams, G.: 1968, 'The Concept of Legal Liberty', in R. S. Summers (ed.), *Essays in Legal Philosophy*, Oxford University Press, Oxford, pp. 121–145.
[22] *Buck v. Bell*, 274 U.S. 200 (1927).
[23] *Jackson v. Indiana*, 406 U.S. 715 (1972).
[24] *Superintendent of Belchertown State School v. Saikewicz*, 370 N.E.2d 417 (1977).

DAVID J. ROTHMAN

WHO SPEAKS FOR THE RETARDED?

In the field of retardation, it appears that we confront a series of issues that never particularly concerned our predecessors. Critical analyses of the perceived worth of the retarded are novel. To ask who speaks for the retarded is also novel. In large measure, these questions were not asked before either because they were not deemed important enough – the retarded were "sports" who did not warrant sustained attention – or more frequently, because the answers seemed self-evident and hence, uninteresting. Quite obviously, this has changed and changed dramatically. The questions are now considered worth asking and the answers are anything but self-evident.

We are, it is true, accustomed in the field of ethics to having a new technology pose new dilemmas. We are the first generation that must concern itself with the implications of genetic engineering or diagnostic innovations like amniocentesis. Yes, precedents and relevant principles do hold, but the technology has at the least altered the agenda of discussion and may even have altered the frame of discussion.[1] To an important degree, technology has had an impact on retardation; because of advances in surgery and neonatology, a number of infants survive who would not have survived in an earlier period. But for the most part, the new questions we must ask do not result just from advances in technology but from critical changes in attitudes. We now examine the worth of the retarded and the right of anyone to represent the retarded mostly because of a deep change in perspective. We see things differently, and it is well worth trying to understand, however briefly, why this is so. What are the origins and the substance of this new outlook?

Perhaps the most useful way to begin to pursue the issue is to examine the altogether too confident answers that our predecessors gave to the question of worth and representation of the retarded. What was it that they had to say that seemed so self-evident to them and, by the same token, much less than obvious to us? For most of this century, certainly since the 1920's, the retarded were defined (and this by the leaders in the field, not the uninformed) as helpless, sick, ill and in need essentially of medical supervision, to be carried out once the natural guardians of the sick, their parents, had given approval. In many ways, this attitude seemed a considerable advance over some earlier definitions of the retarded as menace in terms of their

L. Kopelman and J. C. Moskop (eds.), Ethics and Mental Retardation, 223–233.

corrupting the American stock or the good order of society. As against such primitive and hostile notions, the professional-parent alliance certainly seemed more humane, ready and able to act, in the quintessential Progressive language, in "the best interest" of the retarded.

This confidence in an ability to act in the best interest of a minority group, be it the sick or the poor or the delinquent or the insane, may well have been the hallmark of enlightened and benevolent thought from 1900 at least until the early 1960's. The retarded were only one of many groups (and by no means the most difficult one) to come under this perspective. In all areas, men and women of good will were absolutely confident that their programs would promote the well-being of deprived groups, their confidence encouraged by a sense of the purity of their own motives (they intended to do good and therefore would do good), the promise of American life (which was so great as to make trade-offs unnecessary) and a shared sense of the values that had to be promoted (which made self-doubt disappear). Perhaps the attitude can be most succinctly presented in terms of reformers' antipathy to lawyers and courts, to adversary relationships and formal, fixed and non-discretionary procedures. Thus, when it was necessary to correct the youthful offender (or potential offender), it made no sense whatsoever to bring him before an adult court, equip him with a lawyer, and provide him with all the due process requirements of the Constitution. Instead a special institution, the family or juvenile court, would have a paternalistic judge examine his case and then decide, informally, free of the burdens of jury trials or fifth amendment guarantees, what would be in the best interest of the delinquent. In a phrase, what was good for the delinquent was good for society and this identity of interest guaranteed the right outcome.

One need not dwell in detail on how controversial, really dubious, this judgment now appears to be. Many critics today would strenuously object to notions of one group taking upon itself the prerogative of determining what is in the best interest of another group – whether the case be children, women, students, the mentally ill, or prisoners. There are, it is true, counterarguments (and Supreme Court decisions) that do adopt the more traditional perspective (see, for example, the recent *Parham* decision [6]), but it remains clear that at the least, the debate is fierce, and more, what was once a common assumption is now altogether suspect.

As a society we have certainly become far more prepared to live with fixed procedures and to reduce the discretion of professionals. (One need only read a contemporary consent form given to a hospital admittee or a participant in a scientific experiment to confirm the point.) Unlike our predecessors, we find it appropriate to give those charged with deliquency some (if not all) due process rights, mental patients a right to refuse treatment, prisoners in parole revocation hearings an attorney and, in general, minorities of one or another sort the right to define their own interests. We may not have moved as far from Progressive attitudes as some critics would like, but that we have moved substantially cannot be doubted.

To place this change in the context of retardation, it is now less than persuasive (to understate the point) that the retarded ought to be viewed as entirely helpless or sick; or that it is prudent to leave decisions about their fate entirely to either their parents or their doctors. Certainly when it comes to the borderline or mildly retarded, the position that persons ought to decide for themselves what is in their best interest has powerful support. If the question is, say, marriage or bearing children or using birth control, what might be labelled the post-Progressive judgment is that the retarded should carry the burden of the decision (with all good counselling and advice offered). Indeed, were one addressing only the borderline or mildly retarded it is doubtful that the questions of worth or consent would be especially difficult, any more so than the rights of adolescents. The issues would be framed almost exclusively in terms of decision-making by those with some slightly diminished capacity, which would require attention to defining the boundary lines where autonomy was to be abridged; but in the course of things, such abridgements would be more circumscribed. The balancing would be akin to the "right" of a teenager to request and receive sterilization – a "right" which might well be denied without any fundamental or vital abridgement of autonomy. (At a time when principles of autonomy have so many advocates, some might disagree with this statement and agitate for the right of a teenager to obtain sterilization, but it would be, I am persuaded, a fringe position.) Yes, one can debate whether or not motorcyclists have lost significant autonomy by being required to wear helmets, but it is very difficult even for those who are sceptical of state paternalism to get very excited over the question. As conflicts go, this tends to fall to the less critical side of the spectrum. To apply this to the mildly or borderline retarded, the burden of justification would fall on those who would abridge autonomy. If the case could be made to modify specific rights, we would be dealing with relatively minor rather than wholesale restrictions.

The situation changes notably when we address the severely and profoundly retarded. Here this "easy" answer cannot do. The response to the question of who speaks for the retarded cannot be, simply, the retarded themselves. The handicaps are too great, the disability too deep. One might well wish to glean from facial expression or body language hints of what the the severely or profoundly retarded would prefer as their choice. (And, for example, the hearing officer appointed in the Pennhurst case in Pennsylvania has been doing just that.) But clues are not enough and cannot be determinative. It is evident that some individual or group will have to speak for the severely and profoundly retarded, hence we face the "hard" case: Who should have the authority?

Although an historical analysis does not by its nature supply normative answers, surely the record of past performance has critical relevance to decisions on present policy. In allocating responsibility, the beginning of wisdom is to examine how various actors have fulfilled their responsibilities. Such a survey, let it be clear, is not a very encouraging one. All sorts of people have presumed to speak for the retarded, but their behavior does not justify much confidence.

Let us begin with the *professionals*, the physicians whose credentials and official positions gave them the presumptive right to speak for the retarded. How did they fulfill their duties? The question is quite obviously of large dimension and cannot be answered here in full detail. But one incident may well stand for their performance as a whole. Not that one might not uncover an exception, but rather that the incident points to structural arrangements that make generalizations from it valid.

In the post-World War Two period, institutions for the retarded periodically faced formal accreditation surveys. A body of professionals, including physicians, would tour a given facility and make a report recommending whether or not to accredit it. It was a peer review system, legitimated on the grounds that the professionals would protect the interests of the retarded, that they would speak for them. In research into the early history of Willowbrook State School (an institution on New York City's Staten Island, destined to be the object of an exposé, a court case, and a consent decree ordering that its 5400 residents be returned to the community [5]), I came upon the 1967 report of the "Professional Team of the American Association of Mental Deficiency."[2] By all accounts (including its own) the institution was woefully inadequate. It was desperately overcrowded, understaffed, lacking in programs, in fact lacking in human decency. What, then, was the team's response?

The report opened with a general statement on "Strengths and Weaknesses," The superintendent is "very dynamic and courageous;" Willowbrook is "on the move." Yet, "the institution is plagued by overcrowding and understaffing." Still, there is a promised budget increase. The superintendent "has motivated many of the staff and has received outstanding support from the staff." However, "[t]he turnover rate is very high and only the flagrant offenders are discharged." There is also "some concern about the quality of the people who are employed." Nevertheless, "there is every indication that Willowbrook State School is headed in the right direction." But "it still must be pointed out that ... they have not attained all of their goals due to the inadequacies of overcrowding, understaffing, and complexities of the physical plant."

This combination of double-talk and sugarcoating every disclosure of dreadful inadequacies continued throughout the report. "The nursing staff is quite impressive. The institution has 59 nurses, but 113 vacancies remain." Consider the other possible formulations: Why are there twice as many places open as there are staff on hand? How can 59 nurses possibly care for 5400 residents as handicapped at those at Willowbrook? Should the place be closed down as a health hazard? "The dental department consists of ten people and although this does not meet the AAMD ratio it is remarkable to have a dental staff of this size in an institution for the mentally retarded." Other possible formulations: How can four dentists and six assistants handle the problems of 5400 residents? Should the institution be declared negligent for only having four dentists? "The superintendent has introduced an ambitious plan for remotivation of attendant personnel. This is working very slowly, but effectively.... He has taken some positive steps in an attempt to do something about the excessive absenteeism." Possible variants: What is going on at Willowbrook that staff must be *re*motivated, that absenteeism is excessive, that remedial plans are working *very slowly*, even if somehow or other effectively? "The clothing situation at the institution is considered to be inadequate. The allotment for clothing is insufficient.... The clothing allowances in the dormitories should be reviewed." Or, clients are frequently naked or in rags and tags. "The physical plant is quite inadequate ... another need is to consider the use of more screens throughout the institution." Or, the toilets don't flush, the odors are incredible, the flies abound. For an overall summary, "Willowbrook is an institution which is headed in the right direction.... Although the staff should be commended for their efforts, the level of care, treatment, and training must continue to improve before the standard of care is acceptable." The final potential variant: The standard of care is now unacceptable.

The peculiar nature of this document can be readily explained. The point is not the professionals were unfeeling or uncaring. Rather, they made a political decision of the sort that renders their ability to speak for the retarded suspect; in essence, they identified with the administration, not with the client. The most important fact for them was that the superintendent was trying his best. That he was not succeeding, that the retarded were living in misery, was not his fault, but that of the state or the citizenry or some other unidentified but global force. The team's strategy was to try to give the superintendent leverage in his state capital, to allow him to wave this document to obtain larger appropriations. The strategy did not work; the surveys, here and elsewhere, were misleading in substantive terms and ultimately impotent in political terms. The effort to produce change in this fashion was, for many reasons, ineffective. But still, the type of report that the team gave Willowbrook in 1967 was coin of the realm.

These types of professionals go outside "the system" only with the greatest reluctance. They would never, for example, call in the media, the press or the television cameras to expose the inadequacies that they found. So, too, they would not declare an institution a health hazard, call in the Board of Health, and have it closed down (although a restaurant that in any way approximated the conditions of Willowbrook would not have lasted a day). Why the professionals who visited Willowbrook were so reluctant to take a more daring course can be explained in a variety of ways. Perhaps they made a political miscalculation; perhaps they were cowardly, or too respectful of the reputation of a fellow administrator, or too fearful that he might later seek his revenge when he was on a peer review team for their own institution. But whatever the cause, it is clear that the professionals did not take the retarded to be their primary, let alone exclusive, client. It would seem they were prepared to balance (some would say ignore) the interests of the retarded with the interests of the professionals themselves, or of society; that is, aware of the scarcity of resources, they preferred to see what little money there was go to a more "curable" population, like the mentally ill with a favorable prognosis. In any case, to those familiar with the history of institutions for the retarded, it is difficult to trust to the professionals who ran them and accredited them. For too long, they chose comfortable strategies, not powerful ones.

What of the *parents*? Can they be trusted to speak for the best interest of the retarded? Again, let us remain with the hard case, with the severely and the profoundly retarded who are extraordinarily demanding and who in the past have tended to end up in institutions. Clearly, the parents face an unenviable predicament, for they are rarely in a position to speak for the retarded without deep-rooted conflicts of interest. For one, there are often other children in the family who have rights as well. Will the burden of caring for a retarded child be so onerous as to deprive the others of parental attention? And the parents have their own relationship, too; is the marriage to be sacrificed to the demands of a retarded child? Moreover, when a child is institutionalized, another series of considerations take over that make representing the interests of the retarded all the more difficult. Many parents suffer feelings of guilt at having "abandoned" their child, feelings that will lead (as they often did at Willowbrook) to putting the child altogether out of mind, to not visiting, or to "forgetting" about the child because the matter is too painful. Or it may lead (again, as it usually did at Willowbrook) to an unwillingness to challenge the institutional director for *fear* that he make take reprisals on the child (and transfer him to an even worse ward) or tell them, "If you don't like it here, take your child home." Or, it may produce so much shame that the parents cannot undertake political action to protest substandard conditions because they are too embarrassed by the defect of the child or by their own decision to institutionalize him. In sum, the agenda that parents face is too complex and lengthy to make them "natural" representatives of the interests of their own retarded children.[3]

Beginning in the 1960's, a new group proposed to speak for the retarded, one that heretofore had played practically no role in this field, that is, the attorneys, the public interest lawyers, or the *legal advocates* for the retarded. They came to the assignment by an odd route, from civil rights litigation on behalf of the black minority to prisoner's rights litigation, and then on to mental patients and the retarded, those incarcerated in institutions that could not be easily distinguished from prisons. All were overcrowded, more or less brutal, and desperately short of rehabilitative programs. To the attorneys, interested first and foremost in the rights of their clients, traditional distinctions in the field of retardation (or in mental health) counted for little. Bill of Rights considerations did not distinguish the severely retarded from the mildly retarded. Hence, unlike many others, these attorneys did not presume that the severely retarded should necessarily remain incarcerated (as so many professionals did), or even that they should not have the rights that others did (like the right to reside in the least restrictive alternative). In fact, the attorneys formed an important and effective alliance with the professionals in retardation who were urging that principles of normalization demanded deinstitutionalization; unlike the Scandinavians, who helped to originate the principles, this group insisted that normalization and institionalization were not compatible, that efforts to make a facility over into a "normal" environment were futile, and that in the name of normalization one had to look to return the retarded to the "real" community.

Obviously, a rights perspective and this type of normalization perspective fit well together, and accordingly, efforts to phase down institutions like Willowbrook and put its residents into group homes in ordinary neighborhoods grew more powerful.

However many accomplishments the movement may justifiably cite, the question does remain whether the appropriate answer to the question of who speaks for the retarded is simply the lawyer who heads up the class action suit. It might appear that all that has happened is the substitution of one class of professionals for another, that experts in retardation have now given way to experts in litigation. Indeed, it might be charged that since the lawyers have little detailed knowledge about retardation and are intent on fulfilling a civil liberties agenda (which might not serve the best interests of the retarded), the change does not represent much in terms of well-being. Moreover, there is a significant burn-out rate among attorneys who, unlike many medical professionals, are trained to switch fields, to become "instant experts," which may be fine for the judicial process (where the judge is also a layman), but not for the habilitative.

Well aware of these problems, and a number of others as well, let me suggest that the most appropriate answer to the question of who should speak for the retarded may well be the *formal advocate*, the person who will press for the rights and entitlements of the retarded in an adversarial framework. Although in shorthand, one may end up speaking of the litigator as the advocate, it should be clear that the point is more complicated. The key element is not that the advocate has a license to practice law. Rather, he will press for rights and entitlements before a judicial or administrative body; he will be most comfortable with set rules and unambiguous regulations, preferring to define the authority and obligations of caretakers rather than to trust to their skills or benevolence. In essence, *when one talks of the litigator as advocate one is talking about a context that is formal, adversarial, and regulatory, as opposed to open-ended, paternalistic, and discretionary*.

In part, this judgment can rest on a pragmatic base. Upon reviewing the record of accomplishment, such approaches have managed to reap considerable success. If the criterion be gaining appropriations from legislatures, a court order or consent decree turns out to be more effective than a professional accreditation survey. The amounts of money now devoted to the deinstitutionalization of Willowbrook, for example, far exceed anything that the AAMD was able to accomplish. Moreover, the formal advocate has brought impressive results in client entitlements. Despite a good deal of public confusion that equates all deinstitutionalization with neglect, the retarded have not experienced the kind of abandonment that ex-mental patients all too often suffer. In New York, by executive fiat, the residents of mental hospitals were "dumped" onto the streets, moving from back wards to back alleys. But that has not been true for the court-supervised movement of the retarded. Some 1500 of them now are living in group homes that almost without exception provide decent and humane care, and some of them provide exemplary care.

Indeed, were one to venture a generalization about the outcome of court suits on behalf of the retarded, the results would demonstrate little change in a minority of instances, improvement in more instances, and almost nowhere regression. In other words, suits have not been able everywhere to move a rigid bureacracy, but court-initiated change in the field of retardation has caused considerable good and practically no mischief. Although one can differ about the extent of change, it is clear, for example, that if the Wyatt suit [8] has not promoted a great degree of deinstitutionalization, Partlow is not the hell-hole it once was. Suits in Nebraska and Minnesota have both led to greater use of decent community facilitities; Maine has been slower to progress, but even there some improvements in institutional conditions and ability to place out the retarded have occurred. Even in Pennsylvania, where the Pennhurst case winds its way back and forth through the courts [2, 3, 7], more community living arrangements have been created. By the same token, western Massachusetts now provides the retarded who once resided in Belchertown with far better community residences. Critics are certainly entitled to the argument that court suits appear interminable and do not manage to accomplish all that they promise. Nevertheless, the score card does read quite favorably on the initial outcome of the major class action suits. Rights and needs have moved ahead together, with one not advancing at the expense of the other.

Several qualifications to these rather sweeping statements on outcomes are in order. First, the formal advocate may be helpful in breaking through a tradition of neglect, but it is not certain that over a longer period of time (a decade or two), he will be effective in dictating policy to legislatures. Sooner or later, politics may intrude and legislative efforts to undercut court orders and consent decrees may meet with some success. The very ability of a judiciary to order the expenditure of substantial resources may eventually politicize the issue, and when it comes to politics, neither the court itself nor the litigator have the constituencies necessary to insure victory. Second, a case that demands twenty years of attention may not get it, either because of litigators' flagging attention or scarcity of resources. Public interest law firms are not so well funded that one can anticipate decades of sustained effort.

These problems acknowledged, the formal advocate has more than immediate court victories to commend him. First, this approach creates a forum in which debate, argument, and counter-argument takes place. After all, it is not the attorneys who present testimony but the professionals, caretakers, other experts, and parents as well, so that with some confidence one can note that this system insures that whoever speaks for the retarded will have to do it in open forum, defend the position taken, hear alternative proposals, and respond to them. When difficult decisions must be taken, how much better to arrive at them openly rather than covertly. The likelihood is surely greater than this kind of airing of views will produce a result that is in the best interest of the retarded.

Second, and perhaps even more compelling, formal advocacy necessarily encourages, indeed generates, set procedures and fixed rules. In an area where it is so difficult to allow any one group the right to define the best interests of another, this outcome has decided advantages. Indeed, in other areas where discretion has all too often produced abuses, we also discover an increased reliance upon prior and established rules. The shift in criminal justice from indeterminate to determinate sentences is one notable example. The need for set procedures and fixed rules in the care of the retarded may be even greater. The Willowbrook Consent Decree, for example, has some thirty single-spaced pages of relatively precise instructions. The retarded shall have six hours of programs a day; no more than ten severely retarded persons shall live in a single group home; there shall be two hours of recreation programs a day. To administrators, such explicit standards are cumbersome, rigid, unnecessarily confining and downright annoying. Indeed, they sometimes do introduce a rigidity where greater flexibility might be welcome. We can each think up a case where one or another of these stipulations probably should be waived. But in fact, this specificity has much to commend it. Where discretion has so often produced unsatisfactory results, it may well be better to define the rules of the game and abide by them. Whatever is lost in terms of being able to tailor a response to a particular need is more than offset by the establishment of minmum standards that cannot be violated.

In sum, it is possible to think of answers to the question of who speaks for the retarded that are more persuasive in the individual case. Why should that one extraordinary parent or that one exceptional professional have to operate in a situation where lawyers are omnipresent and rules all-encompassing? But it is not on the basis of these cases that decisions should be made. The area is so conflicted, the history so dark, the predicaments so persistent, that one must think in terms of trade-offs and compromises. And in this context, formal advocacy emerges as the fairest and most effective policy for the retarded.

College of Physicians and Surgeons
Columbia University
New York, New York

NOTES

[1] The analysis that follows rests on the research that Sheila M. Rothman and I carried out under grants from the National Institute of Mental Health and the Field Foundation. The issues raised here will be explored in our forthcoming book (1984), tentatively entitled: *The Willowbrook Wars.*

[2] On file on the Willowbrook Court materials at the New York Civil Liberties Union, New York City.

[3] Those who have followed events in the Becker case [4] will find much evidence there to support this conclusion. By the same token, the tangled efforts of Goldstein *et al.* [1] to justify parental responsibility lead to such unacceptable outcomes that trusting to the parents hardly seems a sufficient answer.

BIBLIOGRAPHY

[1] Goldstein, J., Freud, A., and Solnit, A.J.: 1979, *Before the Best Interests of the Child*, The Free Press, New York.
[2] *Halderman v. Pennhurst State School and Hospital*, 446 F. Supp. 1295 (E.D. Pa. 1977).
[3] *Halderman v. Pennhurst State School and Hospital*, 612 F. 2nd 84 (3rd Cir. 1979).
[4] *In re Phillip B.*, 156 Cal. Rptr. 48 (1st App. Dist., Division 4, 1979).
[5] *New York State Association for Retarded Children v. Rockefeller*, 357 F. Supp. 752 (E.D.N.Y. 1973).
[6] *Parham v. J.R.*, 442 U.S. 584 (1979).
[7] *Pennhurst State School and Hospital v. Halderman*, 451 U.S. 1 (1981).
[8] *Wyatt v. Stickney*, 325 F. Supp. 387 (M.D. Ala. 1971), *modified* 344 F. Supp. 373 (1972).

GERALD L. MORIARTY

COMMENTARY ON DAVID J. ROTHMAN'S "WHO SPEAKS FOR THE RETARDED"

David Rothman emphasized a point upon which I want to elaborate further: that the mentally retarded are a heterogeneous group. I want to argue that he does not think of them as sufficiently heterogenous, and for this reason the problems may be more difficult than he portrays. Although we often speak of diminished intelligence as the core concept defining mental retardation, the case is not that simple. Intelligence is merely a construct, not an actual thing or endowment. Intelligence is a concept employed by psychologists and educators to refer to our ability to accomplish certain standardized test tasks. As such, it is a concept of sharply limited significance. Moreover, measures of intelligence have little direct relation to what we know of the brain's anatomy, function or pathology.

Instead, what a clinician observes are certain specific impairments in cognitive function which are closely correlated with damage to anatomic regions of the brain and to certain of its physiologic or chemical systems. Some examples are: impairments in the use or comprehension of language; defects in the perception of spatial relations; impaired ability to focus attention; failure to perform physical or mental actions in sequence; inability to plan and anticipate consequences; inability to learn or remember; and inability to synthesize perceptions into abstract ideas (for example, one sees only 'men standing there' rather than 'a picture of a baseball game').

Rothman defends the view that, in general, most retarded can speak for themselves. While I agree many can, his position is too facile. *First*, it fails to appreciate how difficult it may be for us to identify with the experiences of some brain-damaged or retarded persons. Many pathologic states probably have no correlate in normal experience. Introspection either cannot inform us about them or can do so only in a distorted way. There are, for example, conditions in which one loses awareness of a side of space and everything in it – not only events occurring in that half of extracorporeal space, but also that half of one's own body. Left and right cease to exist as spatial dimensions. While we can intuitively share the experience of another person who is happy, lonely, perplexed, hopeful or in pain, we cannot genuinely enter into the experience of the person suffering such "hemispatial neglect." And except in a distorted way we cannot understand what our inner experience would be

L. Kopelman and J. C. Moskop (eds.), Ethics and Mental Retardation, 235–242.

if its dimension of symbolic language were to vanish or fail to develop. There are many other examples of ways in which the inner world of persons with brain damage remains mysterious to us. The discrepancies between our experience and theirs could easily lead to enormous problems in making empathic contact, especially with the severely retarded, or prompt us to distort their "perceived worth" in inappropriately monstrous or sentimental ways. We must, therefore, exercise extreme caution in our judgments about what retarded persons feel or what their basic needs are.

Second, Rothman's position fails to appreciate how specific handicaps may constrain one's ability to act autonomously. There are situations, for example, in which a person's mental faculties are fully intact – language, comprehension of social interaction and so on – except for the ability to recall any event which occurred prior to the last minute or two. Learning becomes nearly impossible. The present loses its context. The ability to remember is lost and with it the ability to anticipate consequences. Whereas ordinary human awareness is extended in time, the inner life of such an amnestic individual is not. For these persons, whose structure of consciousness seems to lack past and future but who are otherwise relatively unimpaired, what restraints on liberty if any, are appropriate? How, if at all, should limitations be placed on the social responsibilities and privileges of such individuals?

Finally, using medical rather than psychological descriptions we can be more specific about the differences which separate the mildly, moderately and severely retarded – an important distinction to which Professor Rothman refers. The *mildly* retarded may be difficult to distinguish from normal individuals who are fatigued or placed under stress. They experience impaired concentration, diminished ability to grasp complex situations and lines of action, and difficulty in forming ideas, thinking clearly and using language appropriately. In addition, they may demonstrate an embarrassing proneness to inappropriate social behavior.

The *moderately* retarded may fail to achieve symbolic language, show perceptual handicaps which make it impossible to read and easy to become lost following directions, and have significantly defective memory and diminished powers of manipulating their knowledge to make judgments or anticipate consequences. Their inward lives, their perceptions of their own feelings and needs are more opaque to us – and perhaps to them. These are people who require help, sometimes quite a lot, to accomplish the ordinary tasks of life. But their abilities are sufficiently intact that they can bear all or much of the responsibility for decisions affecting themselves and others, provided support, advice and guidance are available. Professor Rothman's paper intimates that it is our responsibility to provide an entirely new array of services and supports for these persons. Former practices of locking them away with the severely retarded, the psychotic, and the demented, all indiscriminately warehoused in marginally humane or frankly injurious and degrading asylums for the insane, obviously cannot be tolerated. I will suggest that, unfortunately, the alternatives are not so obvious.

The *severely* retarded present the most complex problems in terms of their need for assistance, housing, medical care, and our perception of their rights and worth. Their problems overlap with, but as Professor Rothman pointed out, are significantly unlike those whose impairments are less crippling and more circumscribed. For them the "mental spheres" of thinking, emotion and certain aspects of behavior are severely disorganized. Ability to think and understand may be hopelessly confused, erratic, bizarre or altogether lacking. Memory is usually profoundly defective or absent, language rudimentary or non-existent. The neurological substrate necessary to elaborate anger, pleasure, pain, fear or lust appropriately may be abnormal and the systems necessary to modulate the expressions of those affects may be lost. Behavior may be appallingly self-destructive and dangerous or hopelessly muddled, ineffective, confused and unpredictable. They may display frenetic energy, constant motion and markedly abnormal patterns of wakefulness and sleep. They frequently suffer uncontrollable seizures which can be nearly constant. They may display grotesque uncontrollable movements. Blindness, deafness and disabling physical deformities may compound the problems of their housing and care.

Frequently, severely retarded persons require either physical restraint or constant individual attention. There are metabolic diseases, for example which combine retardation with hyperactivity, an abnormal response to pain, and repeated self-mutilation. One may thus see a child trying with near fanatic persistance to chew off his fingers, bite his tongue and smash his head against walls and floors ([23], p. 1011).

Although I agree with Professor Rothman that the retarded are not "sick" in the all-encompassing categorical sense popular in earlier times, still the severely retarded generally have medical problems of major proportions. Providing for just their medical needs is a formidable task, due in part to the complexity of those needs and in part to the scarcity of demonstrated effective means to meet them. This is true whether one prefers to use large institutions in the country or small ones downtown, but constant nursing supervision and quick access to a physician may be easier to achieve in a larger institutional setting. Devising better alternatives to unacceptable setting such as Willowbrook is far more difficult than closing the institution down.

I turn now to some reflections on the role of medicine in the care of retarded person. Rothman very colorfully illustrated the lack of success that society met historically in trying to turn over complete responsibility for the well-being of the mentally impaired to the medical profession. Are there constructive lessons to be learned?

Historically, the medical profession, like the general public, tended to regard people with mental impairments as comprising a single group with a similar problem. The retarded, demented, psychotic, intoxicated, severely character disordered, sociopathic, neurotic, etc. have often received similar institutional and legal treatment, often been housed together and all regarded as "sick," so that they and their manifold problems fell neatly to the responsibility of medicine.

Although recent efforts have succeeded in large measure in disengaging (some would say disentangling) the retarded from various categories of mental illness, still in important ways our manner of dealing with the retarded reflects confusion about mental illness generally. Partially because we lacked a firm foundation that might have been provided by a thorough understanding of the physiology of cognitive disorders, it became very difficult to distinguish physical illnesses from the outcome of one's social and psychological history as those influenced present habits of thought or behavior. At a time when for a variety of other reasons it was becoming increasingly difficult to specify what one meant by health and illness generally (witness the World Health Organization's over-broad definition of health as "complete physical, mental, and social well being") the medical specialties which deal with "mental health" began to lay claim to expertise at treating a vast array of human problems which could only be regarded as illnesses in a loose or metaphorical sense. The mental health industry opened its doors to the treatment of the "diseases" of quarrelsomeness, bad manners, rudeness, wife beating, nervous tics, gambling, feelings of doom and in fact to nearly every evil to which the human race is subject.

Not surprisingly, many medical specialties and individual physicians with potentially important contributions to make to the understanding of mental impairments and to the care of patients who suffer them came to regard the whole domain of "mental illnesses," including retardation, as too far removed from "real medicine" (from true pathologic physiology) to be of any practical concern to them professionally.

Still other physicians, perhaps encouraged by the broad scope of the conditions considered by so many (the public, third party payers and especially the distributors of vast federal funding programs) to be legitimate medical issues of mental health, found it increasingly natural to make not only medical decisions about the treatment of their patients but decisions about their housing, education, financial support, social life and general well-being. Unfortunately, although the medical profession is reasonably well equipped to study pathologic physiology, to care for persons who are ill and sometimes to influence the course or outcome of disease, it does not include in its professional expertise a mastery of economics, anthropology, sociology, psychology, politics or the administration of large civil service-like institutions. Thus, it is not surprising that physicians demonstrate no special skill as political activists, legal advocates, and/or social workers.

Furthermore, if the profession oversteps the bounds of medicine and assumes too great a share of responsibility, it makes the way easier for others to abdicate their duties or to take no responsibility at all for matters which are broadly societal, economic, educational and political. As a profession, we helped create the illusion that if everyone else left the work in the doctor's hands (and perhaps washed their own hands of it) that the staggering task of accommodating the mentally impaired in our complex society would be taken care of with smooth technical efficiency. The results of this policy are graphically illustrated in Rothman's paper. Professionals alone were not able, or willing, to provide an appropriate level of care for retarded persons. Thus, as Rothman points out, others have assumed the role of advocate.

What, then, should be the physician's role in the care of the retarded? I believe that physicians must work to identify those mental disorders which have a physiological or genetic etiology – the diseases underlying retardation and dementia, the psychotic and major affective disorders, and the systemic and neurologic illnesses which result in impaired cognitive function. Many medical specialties equipped to deal with them (internal medicine, neurology, pediatrics and others) need to renew their enthusiasm for the treatment of those patients. Responsibility for their *medical care* cannot be transferred to social workers, psychologists, psychiatric nurses, occupation therapists, music therapists, dance therapists, etc., who nevertheless contribute so much in the areas of emotional and social rehabilitation. But the problems of many others can almost certainly be better addressed in a setting other than the doctor's office, those whose distress results *not from pathologic physiology* but from social, economic, family or psychological disorganization. Meeting the needs of those persons may be impeded by misrepresenting their problems as "medical" under the rubric of "mental health."

What can one say to Professor Rothman's proposal that the courts may represent an effective, he says the best, forum for sensible mental health and retardation planning? That they can be sometimes effective, I have no doubt. But that their mandated solutions to very complex and problematic technical issues are altogether sensible or beneficial is a point of some dispute.

The experience of the Mental Health Movement illustrates that highly motivated advocates armed with some quite plausible notions about medical care, backed by political power and funded with enormous amounts of federal money may create an impressive furor – and thousands of jobs – but succeed in little more than shuttling their constituents from one dismal situation (the asylum) to another just slightly less dismal (a boarding house downtown or near a suburban shopping center). One ordinarily regards the court and its structurally adversarial proceedings as a last resort to be reluctantly invoked when more effective institutions for planning and problem solving have been exhausted. As an example of why the courts generally come last (or just before open war and bloodshed) rather than first on the list of resources for dealing with life's problems, consider recent legal "solutions" to some mental health problems closely related to retardation: the "deinstitutionalization" of many persons too helpless to care for themselves; the very democratic "right to refuse treatment" by persons painfully confused, delusional and ambivalent but whom treatment might return to clarity of mind [1]; and the statute forbidding involuntary hospitalization unless "treatment" is simultaneously provided – an idea which sounds inspired until it is applied to all those truly incompetent persons for whom no treatment exists but for whose care and support no other option but hospitalization can be found.

Persons with a taste for scientific methodology will recognize in the history of the mental health movement a number of good ideas which unfortunately were not tested before they were enacted and funded, and which turned out to be defective. It is a familiar story in science: an idea sounds thoroughly coherent, clear and convincing when it is proposed, but surprisingly turns out to be thoroughly wrong in practice. For this reason science clings to its slow, laborious and costly methodology of rigorous verification. Legislatures, and to some degree courts, seem to rely on a different methodology in responding to pressure, one characterized in part by the use of persuasion, compromise and razzle dazzle – effective tools for the uses to which they are put, but which quickly entangle their beneficiaries in confusion and frustration when they are applied to the solution of technical problems, such as those posed by the human needs of the mentally impaired. Unfortunately, when dismay takes the place of enthusiasm in those who actually administer court-ordered plans, or whose lives are affected by them, the professional legal advocates have moved on in search of new adversaries. The courts and legislatures may be occupied with fresher, and perhaps better publicized, causes. And in all likelihood they will not be adequately prepared to monitor and adjust the course and impact of their programs as those become clearer with time.

No one would disagree that the courts are an effective means of "breaking through a tradition of neglect," but that is not to regard them as an optimal forum for originating or coordinating societal, institutional and especially technical change. Society – all of us – must still delegate our task of providing for the retarded, as Professor Rothman emphasized in discussions during the conference. And I think most of us agree in feeling dissatisfied with results achieved thus far. But we are learning that our treatment of the retarded has much in common with other contemporary challenges we face as a nation. It urgently requires that we create altogether new ways of working together. In this case the challenge is to create novel institutions and social habits to accommodate the retarded, the psychotic, and the demented in a changing and complex society – and to do so in ways which foster human dignity and freedom for all of us. The format for such moral and social creativity, though, is cooperative and innovative; it is rarely simply adversarial. The dilemmas we face in having to resort to the courts to relieve problems surrounding the institutional care of the mentally impaired should reinforce in our minds the advantages of cooperation over antagonism among those who share responsibility for complex matters. The title of this conference "Natural Abilities and Perceived Worth: Rights, Values and Retarded Persons" and the fact of its occurrence reflect the urgency of this imperative to explore new and cooperative approaches to challenges presented by the mentally retarded in a modern world so complex that over its major conditions perhaps none of us has more than marginal control.

University of Cincinnati Medical Center
Cincinnati, Ohio

BIBLIOGRAPHY

[1] Jackson, D. and Youngner, S.: 1979, 'Patient Autonomy and "Death with Dignity" ', *New England Journal of Medicine* **301**, 404–413.

[2] Stanbury, J.B., Wyngaarden, J.B., and Fredrickson, D.S. (eds.): 1978, *The Metabolic Basis of Inherited Disease*, Fourth Edition, McGraw-Hill, New York.

ARTHUR E. KOPELMAN

DILEMMAS IN THE NEONATAL INTENSIVE CARE UNIT

It is now possible to predict with a high degree of certainty that no matter what care is given, some newborn infants will be severely retarded. Controversies about how best to treat these infants sometimes arise in my area of medicine, intensive care for the newborn. And, to put the issue in perspective, it is important to note that there may be a stark contrast between what one may feel *should* be done and what *can* be done under difficult circumstances and with limited resources. The number of beds or personnel available for neonatal intensive care is sometimes inadequate to serve these patients. How many beds or personnel there are affect the lives or well-being of these sick infants. But this is a funding decision. Ultimately, it is a political one, with which the medical profession has to live.

To be blunt, there are inadequate resources available to provide optimal care for every sick infant. The resources available reflect federal and state funding decisions. Intensive care nurseries often run out of space to admit more infants, but somehow we have always found a place for each infant at some center within our State. Our situation is neither unique nor the worst; sick neonates have on occasion been transferred to us who were born in hospitals several states distant because bed space was unavailable any closer. This is the best we can do under the circumstances, but it is not optimal. Because time is critical for these sick neonates, it would be better and much safer if they could be admitted to an intensive care unit close to where they are born.

Given our limited resources, problems like the following arise: Should an infant who is brain-damaged and we know will be severely retarded, receive prolonged, very costly, life-saving intensive care while other infants with the potential to be "normal" are denied (similar) care because there are no beds available? Is there a difference in the value that should be placed on the life of the retarded as compared with the value placed on the life of the normal newborn infant? To my knowledge physicians have *refused* to make such choices, probably because of their moral or religious convictions, and also for medico-legal reasons.

The parents of brain-damaged or retarded infants, however, sometimes want to make such judgments. They are the people, of course, who must

L. Kopelman and J. C. Moskop (eds.), Ethics and Mental Retardation, 243–245.

provide the day to day and year to year *care* for their child, probably for the remainder of their lives. Over the past 11 years in the practice of neonatology I have worked with many parents who have requested that life-support systems for their critically-ill infant be stopped if there is *very strong* evidence that severe, irreversible brain damage likely to result in retardation has occurred. Is it appropriate for parents to make such requests? Should we, as physicians, honor them, and if so, under what circumstances? [1, 2, 3].

The approach of informing parents, of offering them the option to choose, and of supporting their sometimes painful choices has been advocated by Dr. Raymond Duff at the Yale-New Haven Hospital [1, 2]. He argues that it seems appropriate to allow the parents, with the advice of physicians, to make such a decision. Social policies should be flexible in these cases, he believes, and should allow parents some discretion because they alone will have the lifelong financial responsibility for the medical care and education of the child, and they and their other children will experience the emotional burdens which are devastating to some families. The parents, he argues, also should be most concerned about what the quality of life could be for their infant if he or she survives. Yet, it does not seem right that two newborn infants who have exactly the same problems and potentials may be treated so differently that the result is that one survives and the other is allowed to die.

From personal experience, I know what our local courts' response will be if we ask for help with these decisions: "Never stop. Any quality of life is better than none." The courts and legislators have refused to consider that there are problems of limited resources for neonatal intensive care. Also, is it right for society to insist that tremendous financial and lifelong emotional burdens be borne by the parents and by their other children, without society also assuming a responsibility to help the families of handicapped children financially and emotionally? In many cases adequate or even partial support will *not* be provided.

The last disturbing observation I would like to share with you is that the way things often work out under such difficult circumstances is to the disadvantage of the handicapped or retarded child. It is dangerous to generalize about families, and I hope I will be excused for doing so. I have been impressed that success-oriented, middle-class or upper-class families who greatly value advanced education view the birth of a "damaged" or retarded infant differently than the poor, rural families. To the former, more affluent families, the birth of a retarded infant is often seen as an overwhelming tragedy; they may view the handicap as denying their child any chance of educational or job-related success and the child's tremendous needs as a threat to their own lives and careers. They had planned on having a child but not *this* child. From what I have seen, it seems to me that poor, rural families are much more likely to offer immediate love and acceptance to their handicapped infant, almost regardless of what the child's future holds. Often, I believe, this acceptance has a cultural and/or religious basis.

If these general observations are correct, then we are left with the following irony: Education/success oriented families, likely to have financial resources, are more apt to ask that life-support systems be stopped and their damaged or retarded infant be allowed to die. But poor families, likely to have few financial resources, are more apt to request that no efforts be spared to save their infant regardless of the severity of handicap he or she may have; their child may survive and be loved, but will not have access to badly needed resources for medical, rehabilitative, educational or psychological support. These children require special services to reach their potential. The poor parents cannot afford these services and society (i.e., government) is cutting back on the limited resources which had previously been committed to the care of handicapped children (another policy decision). In short, it seems to me that the most *economically* disadvantaged in our society seem most tolerant and accepting of children with handicaps; but the result is they and their children bear the greatest hardships because our society provides more funds for life-saving intensive care than for follow-up care.

East Carolina University School of Medicine
Greenville, North Carolina, U.S.A.

BIBLIOGRAPHY

[1] Duff, R.S. and Campbell, A.G.M.: 1973, 'Moral and Ethcial Dilemmas in the Special Care Nursery', *New England Journal of Medicine,* 289, 890–894.
[2] Duff, R.S.: 1981, 'Counseling Families and Deciding Care of Severely Defective Children, *Pediatrics* 67, 315–320.
[3] Fost, N.: 1981, 'Counseling Families Who Have a Child With a Severe Congenital Anomaly', *Pediatrics* 67, 321–324.

THEODORE KUSHNICK

HEALTH CARE, NEEDS AND RIGHTS OF RETARDED PERSONS

In my few brief remarks, I should like to address the health care, needs and rights of the mentally retarded, especially the noninstitutionalized retarded, and their families. I will emphasize that the mentally retarded not only have needs and rights, but also obligations to their families. Similarly, families have responsibilities for their mentally retarded members as well as certain rights.

I. THE FAMILY AND THE MENTALLY RETARDED INDIVIDUAL

No mentally retarded person is an isolated individual. Each is a member of a family and as such has certain needs and rights and responsibilities to and for the family, as is true for each other family member, mentally retarded or not. If a family member has a health problem, that person has a responsibility to the family unit to obtain health care, i.e., diagnostic and therapeutic measures. Health problems in any of its members, be it the mother, father, normal children or mentally retarded children, will affect the health of the *entire* family. Even abdication from the family unit by divorce, running away physically or by means of drugs and delinquency, or death, affects the health of the entire family. Therefore, all family members, including mentally retarded persons, should cooperate in appropriate diagnostic and therapeutic efforts to ensure the health of the family unit.

If a child is born with deformities or physical abnormalities indicating a syndrome associated with mental retardation, the child's "obligation" to the family is to participate in diagnostic and treatment efforts. Basically, these investigations are necessary to (1) assist in the future health of the family unit through diagnosis of a syndrome which may recur in future offspring, (2) to assist the parents in gaining knowledge about the causes of the catastrophic event that has destroyed their prior hopes and dreams and shaken their self-esteem as reproductive individuals, and (3) to provide non-specific treatment for the child's illness or handicap. While "no child asks to be born," when he does present himself as a newborn, his parents are responsible for his care and he functions as a family member whose health affects each other member. However, the parents have needs and rights as well as the child.

L. Kopelman and J. C. Moskop (eds.), Ethics and Mental Retardation, 247–251.

II. "RESPONSIBILITY" OF THE DEAD

For the sake of the living and future generations, every effort should be made to understand the causes of each stillbirth or death of a deformed and possibly mentally retarded child. The physicians should not adopt the attitude that "it's for the best," thereby neglecting to recommend complete investigatory procedures. While "heroic" therapeutic efforts might or might not be sought by parents in individual situations, "heroic" diagnostic efforts on the part of the physician are mandatory. These would include detailed physical examination, post-mortem x-ray examination, chromosome analysis, post-mortem photographs, and an autopsy. With such studies, 25% of still-births and newborn deaths do yield evidence of genetic and chromosomal disorders which are important to the future well-being of the family.

Subsequent pregnancies without the knowledge of the previous diagnosis mean a long nine months of anxiety and uncertainty, i.e., "dis-ease" for the parents. In addition, the lack of knowledge with regard to the previous abnormal pregnancy can result in unnecessary feelings of excessive guilt and self-blame, usually felt primarily by the mother; those emotions can persist for a lifetime. If it can be shown that the previous abnormality was the result of an intrauterine infection or a sporadic event without the high risks for recurrence of a genetic disorder, there can be tremendous alleviation of parental emotional disturbance. Given the potential value of such a workup, its expense should not be a serious obstacle; it can be viewed as a health insurance premium for the future life of a family.

III. RIGHTS AND NEEDS OF THE LIVING MENTALLY RETARDED

The rights and needs of mentally retarded individuals are often the same as those of other family members. Besides their legal rights, the loving parental atmosphere and eventual acceptance of these children as individuals with their own sets of abilities is of utmost necessity. The fact is that all but the most profoundly mentally retarded individuals are *emotionally normal*. Although their intelligence level will determine the expression of their emotions, retarded persons can feel just as angry or as happy as any normal person. They cannot express negative feelings with sarcastic remarks and verbal ripostes, but only by physical means, e.g., kicking and striking out.

The basic needs of retarded individuals are the same as those of their normal siblings. In addition, however, they have special needs of at least three kinds: (1) the earliest possible diagnostic and therapeutic efforts, (2) continuous training and intervention programs, and (3) for those moderately and severely mentally retarded, lifelong supervision. Those who are unable to attain an adequate *independent* adjustment in the society in which they live demonstrate the ultimate meaning of mental retardation: the inability to live in an unsupervised and independent fashion in a mentally normal society. Thus, these individuals will not go through high school or college or leave the nest of family life; ongoing care throughout life will be required. Families not only bear the emotional burden, but frequently bear most of the financial burden for this lifelong care. In many states, a family income of over $35,000 per year means that the parents will pay the full bill of $13,000 per year for that individual's care in a state institution. Accordingly, the family should largely decide how to manage their mentally retarded child, inasmuch as the remainder of the community offers varying but incomplete assistance in the child's care. The fact that the family has the primary responsibility for the child does not, however, mean that the community should not intervene to prevent child abuse and/or neglect.

IV. HEALTH CARE AND EDUCATIONAL PROGRAMS

Mentally retarded individuals are often short-changed in their health care because there is a lack of continuity of care on the part of their physicians, as well as an often inadequate knowledge of the means to assist the mentally retarded. The practitioner is generally busy and can offer only acute episodic care for such children. However, this is not because of lack of interest in the patients or their families. The same episodic acute care exists for children with juvenile rheumatoid arthritis or juvenile diabetes mellitus. These are ongoing, chronic disorders that consume an inordinate amount of time for the usual practitioner. Therefore, the best source of ongoing care is in special clinics that are devoted to such specific disorders.

As the parents age, there is greater need for assistance from the community in providing education and training programs, e.g., sheltered workshops, for the mentally retarded. A great deal of lip-service is devoted to the desirability of intermediate care facilities with small homes, minimal supervision and training in home care tasks, such as cleaning and cooking. This is apt to change, however, when an intermediate care facility is attempting to obtain a home near other persons' residences. At that time, the community provides sharp dissent to the idea of having the mentally retarded close to their own homes. The provision of adequate vocational training facilities is also woefully lacking.

One might think of these shortcomings as the community's MR (Moral Retardation). In view of the lack of community support for the retarded, I believe that only the family has the right to decide, by means of techniques such as prenatal diagnosis and abortion, whether they will have or avoid having another mentally retarded child. Prenatal diagnosis is performed because a family *wants* children – not solely for purposes of abortion – but they desire normal children after the previous retarded child. Indeed, only 2%, i.e. 2 of 100, high risk pregnancies subjected to prenatal diagnostic testing will result in the finding of an abnormal fetus and the family's decision to terminate the pregnancy.

V. THE MAJORITY OF THE MENTALLY RETARDED

As Table I shows, only 5% of mental retardation is of genetic or chromosomal causation. Ten to 15% is due to environmental causes impinging on a normal genetic endowment. These two groups are comprised mainly of the moderately to severely retarded, and have parents of normal intelligence who can and do speak in behalf of their children.

Eighty to 85% of mentally retarded children are of the mildly retarded "physiologic" marginal population. *They are not spoken for* because (1) their

TABLE I
Medical categories of mental retardation

Category	%
1. "Physiologic" Polygenic – multifactorial	80–85%
2. Pathologic polygenic – multifactorial – neural tube defect (N.T.D.)	
3. Pathologic Monogenic	
(a) Biochemical – treatment available	
(b) Biochemical – no treatment $\pm$ carrier detection	
(c) Syndromes with other diagnostic markers	5%
(d) Syndromes without diagnostic tests	
4. Chromosome abnormality	
(a) Autosome	
(b) Sex chromosome	
5. Sporadic syndromes	
6. Environmental insults to normal genetic endowment	10–15%
(a) Prenatal – F.A.S., Rubella, Maternal PKU	
(b) Perinatal – Anoxia and CNS hemorrhage	
(c) Postnatal – Intellectual deprivation; emotional deprivation; physical abuse and CNS damage; meningitis – encephalitis; head trauma; toxic encephalopathy	

parents are mainly of the mild mental retardation-borderline intelligence population themselves and are unable to form a vociferous group; and (2) to address the needs of this large group would require a considerable effort. These are individuals who, with proper training, can become independent adults and lose their classification of mental retardation. Programs which can help these children include (1) intervention before pregnancy and during the pregnancy in order to train the mother in parenting skills; (2) continuation of maternally-trained stimulation programs postnatally before the age of 20 months (interventions subsequent to age 20 months are of little value); (3) early-educational programs which are based on the child's abilities and not on the preconceived notion of what an education should be; and (4) the elimination of the concept that vocational education cannot start in early school years, but can only begin after the child has educationally "dropped out" from rigid scholastic regimens that have caused failures and destruction of his self-image. Essentially, the mildly retarded can perform concrete tasks extremely well, and such vocational education is the ticket to their independence as an adult.

Sadly, I believe these efforts will not occur in more than just pilot programs. The reasons are that these parents frequently do not have the emotional, financial or intellectual resources themselves, and the rest of us won't supply them.

East Carolina University School of Medicine
Greenville, North Carolina, U.S.A.

NOTES ON CONTRIBUTORS

Robert M. Adams, Ph.D., is Professor of Philosophy, University of California at Los Angeles, Los Angeles, California.

Larry R. Churchill, Ph.D., is Associate Professor of Social and Administrative Medicine, School of Medicine, University of North Carolina, Chapel Hill, North Carolina.

Cora Diamond, B. Phil, is Professor and Chairman, Department of Philosophy, University of Virginia, Charlottesville, Virginia.

H. Tristram Engelhardt, Jr., Ph.D., M.D., is Professor, Center for Ethics, Medicine and Public Issues, Baylor College of Medicine, Houston, Texas.

Robert L. Holmes, Ph.D., is Professor of Philosophy, University of Rochester, Rochester, New York.

Michael Kindred, J.D., is Professor of Law, Ohio State University, Columbus, Ohio.

Arthur E. Kopelman, M.D., is Professor of Pediatrics, Director of Nurseries, School of Medicine, East Carolina University, Greenville, North Carolina.

Loretta Kopelman, Ph.D., is Associate Professor of Humanities, Director of the Humanities Program, School of Medicine, East Carolina University, Greenville, North Carolina.

Theodore Kushnick, M.D., is Professor of Pediatrics and Director, Divison of Medical Genetics, School of Medicine, East Carolina University, Greenville, North Carolina.

Barbara Baum Levenbook, Ph.D., is Assistant Professor of Philosophy, North Carolina State University, Raleigh, North Carolina.

Laurence B. McCullough, Ph.D., is Assistant Professor and Associate Director, Division of Health and Humanities, Department of Community and Family Medicine, Georgetown University School of Medicine, Washington, D.C.

Joseph Margolis, Ph.D., is Professor of Philosophy, Temple University, Philadelphia, Pennsylvania.

William F. May, Ph.D., is Joseph P. Kennedy Sr. Professor of Christian Ethics, Kennedy Institute of Ethics, Georgetown University, Washington, D.C.

Gerald L. Moriarty, M.D., is Assistant Professor of Neurology, School of Medicine, University of Cincinnati, Cincinnati, Ohio.

John C. Moskop, Ph.D., is Assistant Professor of Humanities, School of Medicine, East Carolina University, Greenville, North Carolina.

Jeffrie G. Murphy, Ph.D., is Professor and Chairman, Department of Philosophy, Arizona State University, Tempe, Arizona.

David J. Rothman, Ph.D., is Bernard Schoenberg Professor of Social Medicine and Director, Center for the Study of Society and Medicine, College of Physicians and Surgeons, Columbia University, New York, New York.

Stuart F. Spicker, Ph.D., is Professor of Community Medicine and Health Care, School of Medicine, University of Connecticut Health Center, Farmington, Connecticut.

Anthony D. Woozley, M.A., is Professor of Philosophy and Law, University of Virginia, Charlottesville, Virginia.

INDEX

PHILOSOPHY AND MEDICINE

Editors:

H. TRISTRAM ENGELHARDT, Jr.

The Center for Ethics, Medicine and Public Issues, Baylor College of Medicine, Houston, Texas, U.S.A.

STUART F. SPICKER

University of Connecticut, School of Medicine, Farmington, Connecticut, U.S.A.

1. H. Tristram Engelhardt, Jr. and Stuart F. Spicker, *Evaluation and Explanation in the Biomedical Sciences* (1975).
2. Stuart F. Spicker and H. Tristram Engelhardt, Jr., *Philosophical Dimensions of the Neuro-medical Sciences* (1976).
3. Stuart F. Spicker and H. Tristram Engelhardt, Jr., *Philosophical Medical Ethics: Its Nature and Significance* (1977).
4. H. Tristram Engelhardt, Jr. and Stuart F. Spicker, *Mental Health: Philosophical Perspectives* (1978).
5. Baruch A. Brody and H. Tristram Engelhardt, Jr., *Mental Illness: Law and Public Policy* (1980).
6. H. Tristram Engelhardt, Jr., Stuart F. Spicker, and Bernard Towers, *Clinical Judgment: A Critical Appraisal* (1979).
7. Stuart F. Spicker, *Organism, Medicine, and Metaphysics* (1978).
8. Earl E. Shelp, *Justice and Health Care* (1981).
9. Stuart F. Spicker, Joseph M. Healey, Jr., and H. Tristram Engelhardt, Jr., *The Law-Medicine Relation: A Philosophical Exploration* (1981).
10. William B. Bondeson, H. Tristram Engelhardt, Jr., Stuart F. Spicker, and Joseph M. White, Jr., *New Knowledge in the Biomedical Sciences* (1982).
11. Earl E. Shelp, *Beneficence and Health Care* (1982).
12. George J. Agich, *Responsibility in Health Care* (1982).
13. William B. Bondeson, H. Tristram Engelhardt, Jr., Stuart F. Spicker, and Daniel H. Winship, *Abortion and the Status of the Fetus* (1983).
14. Earl E. Shelp, *The Clinical Encounter* (1983).